PRINCIPLES OF SCIENTIFIC METHODS

MARK CHANG

AMAG PHARMACEUTICALS, INC,
LEXINGTON, MASSACHUSETTS, USA

CRC Press
Taylor & Francis Group
Boca Raton   London   New York

CRC Press is an imprint of the
Taylor & Francis Group, an **informa** business

A CHAPMAN & HALL BOOK

CRC Press
Taylor & Francis Group
6000 Broken Sound Parkway NW, Suite 300
Boca Raton, FL 33487-2742

© 2015 by Taylor & Francis Group, LLC
CRC Press is an imprint of Taylor & Francis Group, an Informa business

No claim to original U.S. Government works

Printed on acid-free paper
Version Date: 20140602

International Standard Book Number-13: 978-1-4822-3809-9 (Paperback)

Library of Congress Cataloging-in-Publication Data

Chang, Mark, author.
    Principles of scientific methods / Mark Chang.
        pages cm
    Includes bibliographical references and index.
    Summary: "This book focuses on the fundamental principles behind scientific methods. The author uses concrete examples and illustrations to introduce and explain principles. He also uses analogies to connect different methods or problems to arrive at a general principle or common notion. The book explains how the principles of scientific methods are not only applicable to scientific research but also in our daily lives. It shows how the scientific method is used to understand how and why things happen, to make predictions, to present mistakes, and to solve problems. "-- Provided by publisher.
    ISBN 978-1-4822-3809-9 (paperback)
    1. Research--Methodology. 2. Science--Philosophy. I. Title.

Q175.C456 2014
001.4'2--dc23
                                                     2014019443

Visit the Taylor & Francis Web site at
http://www.taylorandfrancis.com

and the CRC Press Web site at
http://www.crcpress.com

Contents

Preface

Our knowledge structures are undergoing dramatic changes as the world further progresses into the Information Age. To not drown in a rising sea of information, knowledge across multiple disciplines becomes increasingly important. *Principles of Scientific Methods* is a book about the fundamental principles or common notions behind scientific methods in different fields.

I wrote this book in order to inspire students to do scientific research, to share experiences and thoughts with experienced researchers, and to stimulate more research into scientific principles. It is challenging for me to systematically write a book on general principles of scientific methods and there will unavoidably be personal opinions. You should read with your critical eyes and suspicious mind. This book is for scientists, researchers, teachers, undergraduates, graduates, and even ambitious high school seniors.

Science as referred to in this book is taken in a broad sense, including natural science, physics, mathematics, statistics, social science, political science, and engineering science. This book focuses on the fundamental principles behind scientific methods. It is not a book about the technical details of scientific methodologies. A principle is often abstract and has broad applicability, while a method is often concrete and specific. We use concrete examples and illustrations to introduce and explain principles and use analogies to connect different methods or problems to arrive at a general principle or a common notion. When I introduce a particular method it is mainly for the purpose of addressing the great idea behind the method, not the method itself. Principles of scientific methods introduced here are applicable not only to scientific research, but also to our daily lives. We study scientific methods for the purpose of understanding how and why things happen, making predictions, and learning how to prevent mistakes and solve problems. Studying the principles of scientific methods is to think about thinking, to enlighten ourselves in scientific research.

Scientific principles are the foundation of scientific methods. They reveal the big ideas behind our scientific discoveries and reflect the fundamental beliefs and wisdoms of our scientists. It is the principles that make the scientific methods coherent. It is the principles, not the methods, that constitute the source of creativity.

There is no doubt that the most important instrument in research must always be the mind of man (Beveridge, 1957). Having said that, scientific methods also play a critical role in the advancement of sciences. According to Huxley (1863), the method of scientific investigation is nothing but the expression of the working mode of the human mind. It is simply the mode in which all natural phenomena are analyzed. Scientific method refers to a body of techniques for investigating phenomena, acquiring new knowledge, or correcting and integrating previous knowledge (Goldhaber and Nieto 2010). To be termed scientific, a method of inquiry must be based on gathering observable, empirical, and measurable evidence subject to specific principles of reasoning (Newton, 1726). Scientific method is a set of procedures consisting of systematic experiment, observation, measurement, and the formulation, testing, and modification of hypotheses (wikipedia.org).

We often talk about the objectivity of science because scientific methods are in general objective, but the most fundamental basis of science is its subjectivity because the principles (such as the similarity principle) supporting the scientific methods are often more subjective than objective. Therefore, science is the conjunction of truth and belief.

It is not my intention to teach students *the* correct principle for each problem. Instead, my intention is to let students know, as early as possible, that there are various principles resulting from different philosophical viewpoints, which may give different, even sometimes conflicting answers to the very same problems. One should be prepared and try to understand why they are different, but one should not get frustrated.

This book includes seven chapters: *Science in Perspective, Formal Reasoning, Experimentation, Scientific Inference, Dynamics of Science, Controversies and Challenges*, and *Case Studies*. With the main goal being to provide the reader with a conceptual understanding of scientific principles and an introduction to several innovative applications, I have included only some basic mathematical/statistical equations in a few sections as necessary. As Professor David Hand (2008) pointed out: "Statistical ideas and methods underlie just about every aspect of modern life. Sometimes the role of statistics is obvious, but often the statistical ideas and tools are hidden in the background. In either case, because of the ubiquity of statistical ideas, it is clearly extremely useful to have some understanding of them."

I have spent time on the figures and diagrams, trying to make these serious, abstract topics a little bit more intuitive and interesting.

We think less than we think we do. An awful lot of what we do is automatic or mindless (Nucci, 2008). Therefore, I hope this book can encourage you to think a little bit more.

Several people have made contributions to the book. First of all, I would like to express my sincere thanks to Dr. Robert Pierce, who always made himself available whenever I asked for help. His careful review and numerous suggestions go beyond normal editorial work. I also want to thank many other reviewers for their time and constructive suggestions. A special thanks goes to Professor Joe Hilbe and Robert Grant for their insightful comments in bringing the manuscript to its current version. Thanks again to David Grubbs, with whom I have worked previously on four books, for his great support and encouragement on this project. Finally I would like to thank my family and friends for their constant support.

Mark Chang (张扬)

Chapter 1

Science in Perspective

1.1 Philosophy of Science

There is no consensus definition of *science*. Scientific realists claim that science aims at truth and that one ought to regard scientific theories as true, approximately true, or likely true. Conversely, a scientific antirealist or instrumentalist argues that science, while aiming to be instrumentally useful, does not aim (or at least does not succeed) at truth, and that we should not regard scientific theories as true (Levin 1984). There is no such thing as philosophy-free science; there is only science whose philosophical baggage is taken on board without examination (Dennett, 1995). It is agreeable that a scientific theory should/can provide predictions about future events; we often take scientific theories to offer explanations for those phenomena that occur regularly or that have demonstrably occurred at least twice.

The world's largest scientific organization, the American Association for Advancement of Science (AAAS) views scientific methodology as a combination of general principles and specialized techniques.

Science is a systematic enterprise that builds, organizes, and shares knowledge in the form of testable explanations and predictions about nearly everything in the Universe. The definition of science evolves. In the early modern era the words *science* and *philosophy* were sometimes used interchangeably. By the 17th century, natural science was considered a separate branch of philosophy. However, *science* continued to be used in a broad sense denoting reliable knowledge about a topic, as used in modern terms such as library science or political science. In modern use, *science* more often refers to a way of pursuing knowledge, not only the knowledge itself. Over the course of the nineteenth century, the word *science* became increasingly associated with the scientific method (research methodology, teaching and learning methods), a disciplined way to study the natural world, including physics, chemistry, geology, and biology.

Scientists share certain basic beliefs and attitudes about what they do and how they view their work. Fundamentally, the various scientific disciplines are alike in their reliance on evidence, the use of hypotheses and theories, the kinds of logic used, and much more. Nevertheless, scientists differ greatly from one another in what phenomena they investigate and in how they go about their work, in the reliance they place on historical data or on experimental findings and on qualitative or quantitative methods, in their recourse to fundamental principles, and in how much they draw on the findings of other sciences. Organizationally, science can be thought of as the collection of all of the different scientific fields, or content disciplines, from anthropology to zoology. There are dozens of such disciplines. With respect to purpose and philosophy, however, all are equally scientific and together make up the same scientific endeavor (AAAS, 1989, 25–26,29).

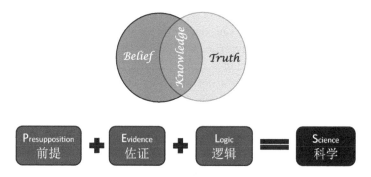

The PEL Model of Science

According to Hugh Gauch (2003), the full force of science's claims results from the joint assertion of Rationality, Truth, Objectivity, and Realism.

Rationality: Rational methods of inquiry use reason and evidence correctly to achieve substantial and specified success in finding truth, and rational actions use rational and true beliefs to guide good actions.

Truth: Truth is a property of a statement, namely, that the statement corresponds with reality. We will discuss more in Section 1.2.

Objectivity: Objective beliefs concern external physical objects; they can be tested and verified so that consensus will emerge among knowledgeable persons; and they do not depend on controversial presuppositions or special worldviews.

Realism: Realism, as regards the physical world, is the philosophical theory that both human thoughts and independent physical objects exist and that human endowments render the physical world substantially intelligible and reliably known. Scientific realism embodies the claim that scientific methods provide rational access to physical reality, generating much objective knowledge.

The PEL model of full use combines **P**resuppositions, **E**vidence, and **L**ogic to reach meaningful conclusions.

Presuppositions are beliefs that are absolutely necessary in order for any of the hypotheses under consideration to be meaningful and true. Every conclusion of science requires presuppositions, just as every conclusion of science requires evidence (Gauch, 2003).

Evidence is relevant data that bear differentially on the credibility of the hypotheses under consideration. Evidence must be meaningful in the view of the available presuppositions.

Logic combines the presuppositional and evidential premises, using valid reasoning such as *induction, abduction,* and *deduction,* to reach a conclusion.

1.2   Theories of Truth

What is truth? What is a proper basis for deciding what may properly be considered true, whether by a single person or an entire society? The topic is so complex that this may not be the proper place for any detailed discussion in this book. But to be aware of controversies is always helpful in scientific research. The theories of truth that are widely shared by published scholars include correspondence, deflationary, constructivist, consensus, and pragmatic theories.

Correspondence theories assume there exists an actual state of affairs and maintain that true beliefs and true statements correspond to the actual state of affairs. Correspondence theories practically operate on the assumption that truth is a matter of accurately copying what has been called "objective reality," and then representing it in thoughts, words, and other symbols (Bradley, 1999).

The *deflationary theory* of truth is really a family of theories that all have a common claim: to assert that a statement is true is just to assert the statement itself. For example, from the redundancy theory, "I smell the scent of violets" has the same content as the sentence "It is true that I smell the scent of violets." So it seems, then, that nothing is added to the thought by ascribing to it the property of truth.

In contrast to correspondence theories, *social constructivism* does not believe truth reflects any external "transcendent" realities. Constructivism views all of our knowledge as "constructed," and truth is constructed by social processes and is historically and culturally specific. Perceptions of truth are viewed as contingent on convention, human perception, and social experience. Representations of physical and biological reality, including race, sexuality, and gender, are socially constructed.

The *consensus theory* holds that truth is whatever is agreed upon, or might come to be agreed upon, by some specified group. Such a group might include all human beings, or a subset thereof consisting of more than one person.

Pragmatic theories hold in common the principle that truth is verified and confirmed by the results of putting one's concepts into practice.

Logically, a truth is what we can't prove wrong, not just what we can prove correct. To prove what is correct or what cannot be proved incorrect, we have to use a certain language or tool of communication. Thus, we use words to define meaning and make arguments, but those words are then further defined by other words, and so on. We finally stop either when we believe the final set of words is clear enough or when we have no time or energy to continue any further!

Intersubjective Agreement:
Criterion for Truth and the Concept of "Fact"

According to Professor Joseph Hilbe (1977), "A statement is true if, taken as proceeding from the objective intersubjectively agreed upon or conventional rules of description, it depicts or properly describes the facts to which the description applies. Facts represent the manner in which our form of life structures the extra-linguistic world and conventionally agreed upon forms of linguistic synonymities according to our socially conceived conceptual framework. Hence, social intersubjective agreement is the criterion of the concept of truth and specifies which statements are true under normal circumstances."

Intersubjective agreement is thus the underlying criterion for both truth and the concept of *fact*. Without further demanding the clarifications of terms "objective... conventional rule," I pretty much agree with Hilbe's statement, since such a view (definition) of truth is operational and very applicable to scientific research.

1.3 Determinism and Free Will

Determinism and Free Will may be one of the most controversial topics in philosophical debate. It is neither my intention nor within my ability to address this comprehensively within the following two pages. However, I believe that it is beneficial to our scientific research to have some basic knowledge about determinism and free will.

Determinism is a broad term with a variety of meanings, as in causal determinism, logical determinism, and biological determinism.

Causal determinism is the thesis that future events are necessitated by past and present events combined with the laws of nature. Imagine an entity that knows all facts about the past and the present, and knows all natural laws that govern the universe. Such an entity may be able to use this knowledge to foresee the future, down to every detail.

Questions about free will and determinism have been debated for thousands of years. It is a profound problem, because many of us believe that without free will there can be no morality, no right and wrong, no good and evil. Were all our behavior predetermined we would have no creativity or choice. However, even some believers in free will agree that in every case our freedom is limited by physical reality and the laws of nature, and therefore free will is limited.

Albert Einstein said: "I do not believe in freedom of will. Schopenhauer's words, 'Man can indeed do what he wants, but he cannot want what he wants,' accompany me in all life situations and console me in my dealings with people, even those that are really painful to me. This recognition of the unfreedom of the will protects me from taking myself and my fellow men too seriously as acting and judging individuals and losing good humour" (Albert Einstein in Mein Glaubensbekenntnis, August 1932).

"Every event is the effect of antecedent events, and these in turn are caused by events antecedent to them, and so on. ... Human actions are no exception to this rule; for, although the causes of some of them are much less well understood than the causes of certain other types of events, it can hardly be denied that they do have causes and that these causes determine their effects with the same certainty and inevitability that are found in every other kind of case. In particular, those human actions usually called free are, each of them, the ultimate and inevitable effects of events occurring long before the agent was born and over which he obviously had no control; and since he could not prevent the existence of the causes, it is clear he could not avoid the occurrence of the effects. Consequently, despite appearances to the contrary, human actions are no more free than the motions of the tides, or the rusting of a piece of iron that is exposed to water and air" (Mates, 1981, p. 59).

Determinism: Chain of Reactions after the Very First Push...

For the purpose of scientific research, a causal relationship between events A and B means two conditions to me: (1) the laws of isolation hold: If A then B; if not A then not B, and (2) the relationship is verifiable, at least in principle, which means that condition 1 is persistent, which further requires that the events in condition 1 are repeatable. For a finding to become a scientific law, it must be verified independently. The repetition of events is what gives causality its great usefulness: we use it to make predictions.

There can be other definitions of what constitutes a causal relationship. For example, one may say: "God created the universe," which means that with God we have the universe and without God we would not have the universe. However, we cannot in principle prove or disprove the statement, so it can be considered as a logical truth. We will have a brief discussion of such logical truths in Chapter 2. In this book, we are mainly interested in those causal relationships that either can be verified in principle over time through repeated events or are provable in principle through logical deduction (Chapter 2).

Free will believers also consider that there are causal relationships, but their pursuit of causes stops right there at free will. To them the cause of an event (e.g., making choices) can be scientifically based or by way of an agent's free will. If it is due to free will, we should not ask further what causes free will to make such a choice or why different people have different free wills. To the believers, free will is the origin of many outcomes, just as many of us believe the big bang is the origin of time.

However, it will not be scientifically interesting to say, "It happens because of free will." We, as scientific researchers, always ask why we make, or what causes us to make, certain choices. I often ask myself: With the ultimate development of neuroscience and understanding of the human brain, how much space remains for free will? On the other hand, we may be equally terrified by determinism: "Everything happens for a reason, including every piece of our thoughts, every one of our choices, our beliefs and emotions, every tiny movement we make or action we take...." Humans and their actions are machine-like, according to a strictly deterministic view.

In any theory we pursue maximal consistency, but complete consistency is not always possible. The conception of the universe is a simplification of the universe since our skull-encased brain is much smaller than the universe; this is just like putting a huge vase into a tiny vase, where what's inside the smaller vase are the broken pieces of the larger one. With natural laws we hope these broken pieces can in principle be put back together to restore the original universe. Each human skull holds the universe that includes itself and all other skulls. Is it logically possible?! Similarly, we think about thinking and about thinking about thinking. With such high dimensions in cognitive space, the consistency in Aristotle's logic isn't always possible.

It is true that the universe is or can be deterministic even if not everything happens for a reason. I like to think (even if I would not try to prove it) that there is a single history, and everything is unique over its course. When we slice history into pieces and judge two pieces to be the same, we intentionally or unintentionally ignore their differences; at least we ignore the different "neighboring pieces" in the chain of history. Such slicing and grouping similar pieces together artificially create repetitions of that we may consider "the same,"and allow us to "discover" natural laws or causalities. In this sense, a natural law is first invention (grouping the similar pieces) and then discovery (finding the pattern). If you don't agree with this view at the moment, you might change your mind after finishing the next section: The Similarity Principle.

The grouping that our minds carry out in hand with the similarity principle makes everything happen for a reason, but at the same time, due to the approximation of such grouping, every law or causal relationship has exceptions.

1.4   The Similarity Principle

According to Albert Einstein, the grand aim of all science is to explain the greatest number of empirical facts by logical deduction from the smallest number of hypotheses or axioms.

Science is a study of the reoccurrences of events. Its purpose is to reveal or interpret the "causal relationship" between paired events. One of them is called "cause" and the other is called "consequence" or "effect." The reason we study history is to develop or discover the law that can be used to predict the future when the cause repeats. An event in a causal relationship can be a composite event. However, science does not simply ignore all single occurrences. Instead, one-time events are grouped into "same events" based on their similarities. In such a way we artificially construct reoccurrences of events.

According to Karl Popper (2002), any hypothesis that does not make testable predictions is simply not science. A scientific theory is constantly tested against new observations. A new paradigm may be chosen because it does a better job of solving scientific problems than the old one. However, it is not possible for scientists to test every incidence of an action and find a reaction. You may think the brain is so amazing because it can do science, but that is also a limitation of our brains—they are unable to deal with each event individually.

A cause–effect pair is the observation of a repetitive conjunction of particular events for which we postulate their linkage (causal relationship). In this philosophical sense, causality is a belief, an interpretation of what is happening, not a fact but at most a thing that cannot be verified or whose verification rests on the basis of another belief. There is a single, growing complex history. To simplify it, people try to identify the patterns or "repetitive conjunctions of particular events" and deduce "laws" or causal relationships from the process. The term *cause* relies on the recurrence of the events and the definition of *same* or *identical*. The recurrence here is in regard to the recurrence of the "cause" and the "effect," both together and separately. Consider the law of factor isolation: If factors A and B exist, fact C exists and if we eliminate B, C disappears, then factor B is a cause of fact C. However, there is no such exact reoccurrence. The reoccurrence of an event is only an approximation in the real world; we ignore certain distinctive details, and therefore it is always somewhat subjective.

The notion behind prediction is causality, whereas the notion of causality is the *similarity principle*: similar situations will likely result in the same outcome. When we say "the same," we ignore the differences that might be hidden and contradict the scientific law that is held. Once such a contradiction is uncovered, an exception to the law is observed, which may call for some modification of the law. Both Science and Statistics acknowledge the existence of exceptions, but the former doesn't act until an exception occurs, while the latter acts when the unexpected is expected.

Most of us would agree that the goal of scientific inference is to pursue the truths or laws that approximate causal relationships so that our brains can handle them. Every pair of things (events) are similar in one way or

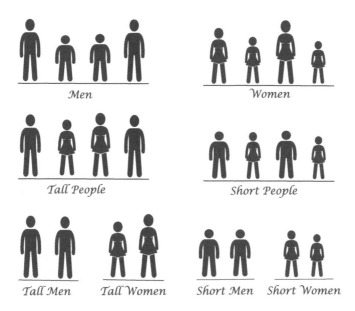

Men

Women

Tall People

Short People

Tall Men Tall Women Short Men Short Women

Similarity Grouping

another, so we can see the reoccurrences of the "same event" through grouping, based on similarities and ignoring the differences. Importantly, such grouping is necessary for our brains to efficiently handle the reality. It is this grouping that lets us see the repetition of things (both reasons and outcomes). Only under reoccurrences of "same events" can a causal relationship make sense. Such implicit grouping of similar things often creates intangible controversies in scientific inference. The differences we ignore are unspecified or hidden; only the similarities are known and specified. The ignored differences cause a difference in defining the sets of "same things," called the "causal space," which will be discussed further in Chapter 4. In other words, what things are considered similar is somewhat subjective. Different people have different opinions for different questions. If you are a tall man, from which of the following four groups do you infer your characteristics (e.g., life-expectancy, the likelihood of being successful in career, or the chance of having cancer): men, tall people, tall men, or the entire population?

We all use the similarity principle nearly every moment throughout our lifetimes. It says: A similar condition will lead to a similar consequence. Without the similarity principle, there will be no analogy and no advancement of science. But how do we define two things as being similar? We can say two things are similar if they have similar consequences. But this is really nothing but inversely stating the similarity principle.

1.5   The Parsimony Principle

William of Occam (or Ockham) (1284–1347) was an English philosopher and theologian. His work on knowledge, logic, and scientific inquiry played a major role in the transition from medieval to modern thought. Occam stressed the Aristotelian principle that entities must not be multiplied beyond what is necessary. This principle became known as Occam's Razor or the *parsimony principle*: The simplest theory that fits the facts of a problem is the one that should be selected. However, Occam's Razor is not considered an irrefutable principle of logic, and certainly not a scientific result. According to Albert Einstein, "The supreme goal of all theory is to make the irreducible basic elements as simple and as few as possible without having to surrender the adequate representation of a single datum of experience."

According to Douglas Hofstadter (Mitchell, 2009, Preface), the concept of reductionism is the most natural thing in the world to grasp. It is simply the belief that "a whole can be understood completely if you understand its parts, and the nature of their 'sum'". Reductionism has been the dominant approach since the 1600s. René Descartes, one of reductionism's earliest proponents, described his own scientific method thus: "... to divide all the difficulties under examination into as many parts as possible, and as many as are required to solve them in the best way ... to conduct my thoughts in a given order, beginning with the simplest and most easily understood objects, and gradually ascending, as it were step by step, to the knowledge of the most complex."

As Gauch (2003) pointed out, parsimony is an important principle of the scientific method for two reasons. First and most fundamentally, it is important because the entire scientific enterprise has never produced, and never will produce, a single conclusion without invoking parsimony. Parsimony is absolutely essential and pervasive. Second and more practically, parsimonious models of scientific data can facilitate insight, improve accuracy, and increase efficiency. Remarkably, parsimonious models can be more accurate than their data.

In mathematical and statistical modeling, the *Parsimony Principle* can be stated: from among models fitting the data equally well, the simplest model should be chosen. The goodness of fitting is judged by predictive accuracy, explanatory power, testability, fruitfulness in generating new insights and knowledge, coherence with other scientific and philosophical beliefs, and repeatability of results. A simpler model is often less precise for certain problems but applicable to more problems.

We use the following figure to illustrate simple versus complicated models. The current data involve the effect of variable x and random noise. Suppose we fit a curve passing all data points (circular dots) using a complicated function

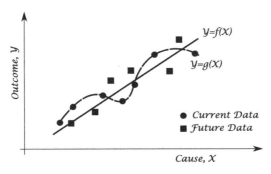

Parsimony Principle

$g(x)$. This function fits current data well, but it might fit the future data poorly because noise will change every time we get new data (see the squared dots). On the other hand, if we use a simpler function $f(x)$ to approximate the current data points, minimizing the effect of random errors, the function $f(x)$ will fit reasonably well next time since it reflects mainly the effect of x. In other words, from a predictive point of view, the simpler model $f(x)$ will do a better job in predicting the squared points than the more complex model $g(x)$.

It is interesting that the parsimony principle is also sometimes the key to a successful invention as we have seen in the huge success of iPhone applications— simplicity rather than comprehensiveness or sophistication. However, parsimony does not always work. Consider the following consequence of evolution. Humans are very robust, i.e., there are many redundant parts that may share similar responsibilities or functions, so that when one fails to perform the job, another part is there to take its place.

1.6 Essence of Understanding

In ancient China, there were two scholars, Zhuangzi and Huizi, who argued about "knowing." Zhuangzi said: *"Look how happy the fish are!"* Huizi asked: *"You are not a fish, how do you know they are happy?"* Zhuangzi answered: *"You are not I, how do you know I don't know?"* Such arguments can go on and on since scholars agree that one can only be understood by oneself. But what is the meaning of *understanding* (or the meaning of *meaning*) anyway? If an artificial intelligence (computer) can generate the text (or say) "$1+1=2$" or "It is going to rain tomorrow," do you think the agent understands what the computer said? Do we know what, if anything, the computer meant by the sentence?

A concept or term is defined by other concepts and terms, which are further defined by other words and terms, and so on. Thus, it is believed that a very initial set of words is necessary in communication. This first set of words is used to explain other words or concepts; more words can then be explained further by those words that were just explained, and so on. The meaning of each word is different for different individuals and changes over time. People often think they have understood each other when they establish a mapping (or agreement) between what they have perceived. It is basically the situation: "I know what you are talking about" or "I understand what you mean." What I think your understanding is and what you actually understood can be completely different. But it does not matter as long as I think I understand that you understand, and vice versa.

Imagine that we connect words by arrows to form a network, called a *wordsnet*. Let's consider the wordsnet of an English dictionary. For example, starting with the word *mother*, which is defined as "a female parent" in the Webster dictionary, we use three arrows to connect "mother" to the three words *a*, *female*, and *parent*. Then these three words are further defined by other words and linked by arrows, and so on. In this way, we are constructing a wordsnet that is a cyclic network with many loops. Similarly, the wordsnet of a human brain is also a cyclic network with many loops. Furthermore, the conceptsnet (a similar network connecting concepts) of a human brain is a cyclic network with loops. For example, understanding is explained by comprehension and comprehension is inversely explained by understanding, in the Merriam-Webster online dictionary at http://www.merriam-webster.com/dictionary/. Another example from the same dictionary, starting the search with the word *confused* we get "being perplexed or disconcerted." Then, using *perplexed* for further search, we get "filled with uncertainty: puzzled." Continuing, we use *puzzled* to get "challenged mentally," and search *challenge* to get "presented with difficulties." Finally, if we search *difficult*, we find "hard to understand: puzzling."

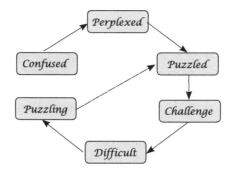

Essence of Understanding: Words-Mapping

You see, the word *puzzled* is explained by "puzzling." But after going through many such loops (often much bigger loops), people feel they are getting the meaning of the word or concept.

Such concept-mapping in the process of understanding also occurs when we learn how to make any machine, especially when we imagine the machinary-making in our minds.

How do we explain the ability of a blind person to discuss color or vision with a vision-perfect person? That is exactly because "understanding" is nothing but a words-matching game!

Metacognition is special understanding process, for instance, thinking about thinking. One of the hardest problems is trying to understand the mind while using it as a tool. This requires us to use a tool to define a tool or a process to define a process.

1.7 Discovery or Invention

Is science a *discovery* or an *invention*? Is it in the external world (outside the human mind) and so must be discovered, or does it lie in the mind and is therefore invented?

Can an artificial intelligent (AI) discover new knowledge? It depends on how we define the term *discover*. For instance, can an AI discover the theory of evolution? What does the term *discover* mean here? If an AI did generate the exact same text as Darwin's evolution theory but before Darwin did, would it mean that the AI has "discovered" the theory? And since a "discovery" is presumably made by its "discoverer," would this be the AI or the human reader? Should we explain the text differently from Darwin's text just because it was generated by a machine? If we do explain the machine-generated text the same way as we did Darwin's text, should we call it a discovery? If we do, we in fact have created an AI that can carry out scientific discovery because such an AI is just a random text generator.

When a machine can be made "human" is not only dependent on the advancement of computer (AI) technology, but also dependent on how we change our view, including moral standards, of the concept of "human," and on the extent to which we can accept the machine-race human as human without racial discrimination.

Since determination of a discovery or invention is dependent on whether it initially exists outside of a human mind, it is critical to clarify the connotation and denotation of a human or identity. Puzzles about identity and persistence can be stated as: under what conditions does an object persist through time as one and the same object? If the world contains things which endure and retain their identity in spite of undergoing alteration, then somehow those things must persist through changes (Washington.edu)?

We replace malfunctioning organs with healthy ones. We do physical exercises to improve our health. We like to learn new knowledge, the quicker,

Discovery and Invention

the better. We try hard to forget sad memories as quickly as we can. As these processes continue, when does the person lose his identity?

Suppose Professor Lee is getting old. He expressed his wish to have a younger and healthy body, whereas a healthy young student, John, truly admires Professor Lee's knowledge. After they learned of each other's wishes, they decided to switch their bodies or their knowledge, however you wish to say it. In the operation room, the professor's knowledge (information) was removed from his brain and transferred to John's brain. At the same time whatever original information that was in John's brain was removed and transferred to Doctor Lee's brain. The operation was carried out using the "incredible machine." Now the question is: Who is who after the operation? Would they both be happy with the operation? Please think, and then think again before you answer.

If we cannot clearly define what is a person, how can we make clear differentiation between discovery and invention at fundamental level?

1.8   Observation

Philosophically *observation* is the process of filtering sensory information through the thought process. Input is received via hearing, sight, smell, taste, or touch and then analyzed through either rational or irrational thought. Therefore, observing is a part of the processes of learning.

In discussing the thoroughly unreliable nature of eye-witness observation of everyday events, W. H. George says: "What is observed depends on who is looking." He tells the following story (Beveridge, 1957):

> At a congress on psychology at Gottingen, during one of
> the meetings, a man suddenly rushed into the room chased
> by another with a revolver. After a scuffle in the middle of

the room a shot was fired and both men rushed out again about twenty seconds after having entered. Immediately the chairman asked those present to write down an account of what they had seen.

Although the observers did not know it at the time, the incident had been previously arranged, rehearsed and photographed. Of the forty reports presented, only one had less than 20 per cent mistakes about the principal facts, 14 had from 20 to 40 per cent mistakes, and 25 had more than 40 per cent mistakes. The most noteworthy feature was that in over half the accounts, 10 per cent or more of the details were pure inventions.

Such an error in registering and reporting one's observation has its origin in the mind itself since the mind has a trick of unconsciously filling in gaps according to past experience, knowledge, and conscious expectations. We see only what we know. We are prone to see what lies behind our eyes rather than what appears before them. We often see something repeatedly without registering it mentally.

Beveridge (1957) further pointed out: "It is important to realize that observation is much more than merely seeing something; it also involves a mental process. In all observations there are two elements: (1) the sense-perceptual element and (2) the mental, which, as we have seen, may be partly conscious and partly unconscious."

A scientific experiment is often purposely arranged to isolate certain events which are of interest to the experimenter and are observable by the aid of appropriate instruments. What is often more difficult is to observe (in this instance mainly a mental process) resemblances or correlations between things that on the surface appeared quite unrelated.

Our mind is selective. One cannot observe everything closely; therefore, one must discriminate and try to select the significant. The "trained" observer deliberately looks for specific things which his training has taught him are significant, but in research he often has to rely on his own discrimination, guided only by his general scientific knowledge, judgment, and perhaps a hypothesis which he entertains. Trained scientists always expect (lookout for) the unexpected (Beveridge, 1957). Observing as discussed here does not necessarily mean "seeing," but can also be feeling or sensing with any of the other senses. What follows next is a story illustrating how a great scientist is always ready to expect (look out for) the unexpected.

Archimedes (287–212 BC) was one of the greatest mathematicians and inventors of all time. He discovered fundamental theorems concerning the center of gravity of plane figures and solids. His most famous theorem gives the weight of a body immersed in a liquid, the Archimedes Principle: A body immersed in a fluid is subject to an upward force (buoyancy) equal in magnitude to the weight of fluid it displaces.

It is said that Archimedes discovered the principle of displacement while stepping into a full bath and realizing he felt less pressure from the bottom of the bathtub when there was water than when there was not. He realized that the water that ran over equaled in volume the submerged part of his body. Through further experiments, he deduced the principle. As legend tells it, Archimedes was so excited with his discovery that he hopped out of the bath and rushed naked into the street yelling triumphantly, "Eureka! Eureka!" (Greek for "I have found it!").

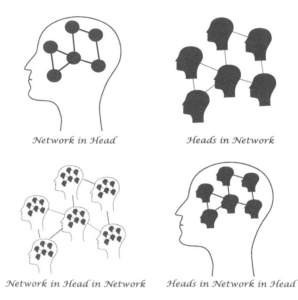

Network in Head   *Heads in Network*

Network in Head in Network   *Heads in Network in Head*

Observability in Question: Recursively Self-Inclusive

Interestingly, there is still room for question about observability: We have the concept of a network (cognitive space: stuff in head), of ourselves being in the network (social space: heads in stuff), and recursively, of stuff in heads in stuff as illustrated in the figure above. Can one really observe oneself?

1.9   Experimentation

Experimentation is the most commonly used tool for scientific research. The main difference between experiments and observational studies is that in observational studies hypotheses are tested by the collection of information from phenomena which occur naturally, whereas an experiment usually consists of making an event occur under known conditions where as many extraneous influences as possible are eliminated and close observation is possible so that relationships between phenomena can be revealed. This is due to the universal principle of factor isolation as described in Section 1.4: If factors A and B exist, fact C exists and if eliminate B, C disappears, then factor B is a cause of fact C (given the fact that A always exists). Here C can be a composite factor.

There are vast varieties of experiments in different subject fields and different types of experiments. There is a distinction between prospective and retrospective experiments. A retrospective study poses a question and looks back. The outcome of interest has already occurred (or not) by the time the study is started. For instance, we may be interested in how cigarette smoking relates to lung cancer. To this end, we can look into the patients' hospital records about their smoking history and lung cancer.

In contrast, a prospective study asks a question and looks forward. The studies are designed before any information is collected. For example, in a clinical trial studying the effect of a cancer treatment, patients are identified to meet a list of prespecified inclusion and exclusion criteria. The qualified subjects will be randomly assigned to receive one of two treatments. The randomization here has a clear purpose, which is to reduce the imbalance of confounding factors (factors that are related to both the treatments and outcome, but are not causal factors) between the two treatment groups. This is done so that when we observe the difference in the outcome (survival time in this case), we can conclude it is because of different treatment effects rather than other factors. To remove or reduce the confounding effects, two groups are usually required in an experiment with one of them serving as the "control." An experiment with a control group is called a controlled experiment.

There are also uncontrolled experiments, in which no randomization is usually applied and the conclusions reached may not be as robust as those from controlled experiments. Controlled experiments are broadly recognized as the gold standard for experimentation in scientific research. The two groups in a controlled experiment do not have to be different subjects. For instance, in a crossover experimental design, every subject is given two different treatments sequentially (e.g., treated with drug A is the first week and B in the second week). In such an experiment, each subject serves as his/her own control.

We should know that there is a subjective component that may dramatically affect the outcome, i.e., bias. The known placebo effect is a major one. The outcome may be affected by the knowledge of what has been taken by the subjects. For instance, a patient known to be taking an active drug may inflate his/her evaluation of improvement, whereas a patient known to be taking a placebo may deflate his/her evaluation of improvement. For this reason, researchers invented the so-called blind experiment, in which the people involved in the experiment do not know what treatment is being given to which subject. Such blinding mechanism can be applied, for example, to patients only (single blinding), or to patients and physicians (doubleblinding).

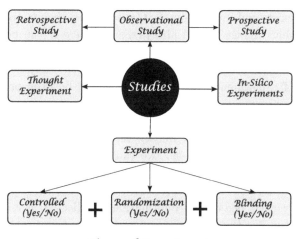

Types of Experiments

An *in-silico experiment* (computer simulation or Monte Carlo simulation) is an experiment performed on a computer. Due to rapid developments in computer science and engineering in the past two decades, many experiments can be done on computer. There are emerging fields such as Monte Carlo simulation, computational chemistry, computational biology, molecular design, experimental math, and others.

An in-silico experiment can be combined with a traditional experiment to accelerate research and reduce the cost. We will see an application of an in-silico experiment for molecular design as an example in Chapter 7.

Yet another type of experiment is the thought experiment (TE). *Thought experiments* are carried out in one's mind instead of in a laboratory. The purpose of a TE can be to disconfirm a theory by disclosing a conflict between one's existing concepts and nature. It can also be for constructive purposes or it plays both a destructive and constructive role (Brown, 1991; Kuhn, 1977). Thought experiments typically feature the visualizing of some situation, virtually carrying out an operation, "observing" what happens, and drawing a conclusion. They can also illustrate the fallibility of a conclusion.

One of the most beautiful early examples of a thought experiment attempts to show that space is infinite (http://plato.stanford.edu): If there is a purported boundary to the universe, we can toss a spear at it. If the spear flies through, it isn't a boundary after all; if the spear bounces back, then there must be something beyond the supposed edge of space, a cosmic wall that stopped the spear, a wall that is itself in space. Either way, there is no edge of the universe; space is infinite.

1.10   Interpretation

After data are collected, we need to interpret them through statistical analysis. *Analysis* is, on one hand, the activity of breaking an observation or theory down into simpler concepts in order to understand it; on the other hand, it integrates piece-wise information to reach meaningful conclusions. The main objective of analysis is to find a causal relationship. According to nineteenth-century philosophers, a *causal relationship* exists (1) if the cause precedes the effect; (2) whenever the cause is present, the effect occurs; and if the cause is the only cause then (3) the cause must be present for the effect to occur.

We should clearly differentiate the "truth" from "logical truths." There are often many logically possible interpretations (logical truths). A *logical truth* means that a proposed causal relationship sounds logical but may not be the truth at all.

We should also be aware that common sense and knowledge of specific fields are necessary to sensibly interpret the results of an experiment. An *association* is not a causal relationship. For instance, data show that people who carry matches are more likely to have lung cancer than people who do not carry the matches. However, that is because smoking cigarettes can cause cancer and smokers usually carry matches. Thus, we call smoking a confounder. To make the thing even more complicated, because secondary smokers are more likely to have lung cancer than smokers, when one compares these two groups, smoking is a safe factor to reduce lung cancer. In statistics, a confounder is an extraneous variable that correlates positively or negatively with both the dependent variable and the independent variable.

Using experimental data in an isolated way or fragmentally will undermine the integrity of scientific research. For instance, we often intentionally or unintentionally screen the observations and select ones that fit or support our preformed conclusions. This is easily seen in religious societies, where certain types of stories are told many times, while the stories supporting the opposite argument are often neglected. Such a tendency is called *confirmation bias*. Our brain is very selective: a person with a positive personality hears the more positive things, while a pessimistic person is attuned to more negative things.

Conformation bias is most often seen in scientific publications, where positive results are much easier to publish than negative results. The confirmation bias and the so-called multiple testing issues can make scientific discovery

Interpretation: It Is Necessary to Cook before Serving.

really controversial. Confirmation bias can occur when (1) we collect only the events that fit preformed conclusions and (2) data are collected until we reach the desired conclusion.

Suppose we want to prove that there is more likely to be heads than tails in coin-tossing experiments. First, we cannot just count only the events with heads to prove our conclusion; neither can we put more effort into collecting data from all the experiments in which the proportion of heads was more than that for tails. Second, we cannot keep tossing the coin until there are more heads than tails. Because of the randomness in an experiment, there can be more than 50% or less than 50% heads observed at some timepoint in an experiment.

In statistics, *hypothesis testing* is used to reduce the false positive claim. For example, if we want to test if there is a greater than 50% probability of heads in the coin-tossing experiment, we can construct the hypothesis test as follows:

Null hypothesis: 50% probability of heads

Alternative hypothesis: a greater than 50% probability of heads

We always put what needs to be proved in the alternative hypothesis. To claim the conclusion (heads >50%), we have to reject the null hypothesis and ensure that the false positive error rate is below a nominal value α, called the level of significance for the test. The α is usually a small positive decimal, e.g., 5%. If the null hypothesis is rejected, the false positive rate is no more than 5%. In the above coin-tossing experiment, suppose we decide to toss exactly 100 times before the experiment begins. If the outcome shows 59 or more heads, we reject the null hypothesis and conclude there is a greater than 50%

probability of heads in the coin-tossing experiment. This is because when the null hypothesis is true (50% probability of heads), there is only 5% probability of getting 59 or more heads—or the error rate is 5%. Alternatively, if we decide to toss the coin 10 times, then if at least 9 heads occur in 10 tosses we reject the null hypothesis; otherwise we don't reject the null hypothesis. In such a way we can also control the false positive error rate at the $\alpha = 5\%$ level.

However, things are not that simple. There is a very controversial multiple testing issue. It concerns the inflation of the false positive error rate when many hypotheses are tested in the same experiment or the same hypothesis is repeatedly tested with different experiments. Model selection without multiplicity adjustment is a postdictive approach and inflates the error rate.

There are many other common issues in data analysis and interpretation, such as regression to the mean, Simpson's paradox, bias, confounding, the placebo effect, and internal and external validity in general. We will discuss them later.

1.11 Qualitative and Quantitative Research

Qualitative research studies the *why* and *how* of phenomena, while qualitative research aims at answering questions beginning with words such as *when*, *where*, *how many*, or *how often*. Qualitative analysis is mainly descriptive and often imbued with words such as *because, good, fair, bad, often, rarely*, while *quantitative research* nails the answer with numbers.

Qualitative research is exploratory in nature. It is often used when we don't know what to expect, to better-define a problem or for the purpose of developing an approach to a problem. It is also used to go deeper into issues of interest. In general, qualitative research generates rich, detailed data that contribute to in-depth understanding of the context.

Qualitative and quantitative studies usually provide different answers to problems. For instance, obesity results from energy imbalance: too many calories in, too few calories burned. A qualitative research result may state that exercise can reduce obesity, while quantitative research may conclude that two hours of exercise per day can reduce the chance of obesity by 20%.

A commonly used qualitative method in social science is the so-called focus group, in which a group of subjects are involved in in-depth interviews, often with open-ended questions. Coding is an interpretive technique that organizes the data and provides a way to introduce the quantified interpretations. However, such coding should be used with caution because of the limitations of the source data and because the choice of coding method is somewhat arbitrary.

Qualitative analysis seeks to discover patterns such as changes over time and/or associations or possible causal links between variables, providing a foundation for further research using quantitative methods.

As we all know, causation is different from association. Just because a person is found holding a gun at a crime scene does not mean he is a killer. Causation must be an association, while an association is not necessarily causation. In qualitative analysis, to determine if an association is potentially causal the well-known Bradford Hill criteria are frequently useful. *Hill's criteria* for causation were proposed by English epidemiologist Sir Austin Bradford Hill in 1965.

In his paper Hill (1965), stated: "Our observations reveal an association between two variables, perfectly clear-cut and beyond what we would care to attribute to the play of chance. What aspects of that association should we especially consider before deciding that the most likely interpretation of it is causation?" He then discussed a list of nine criteria with illustrative examples. Hill developed the criteria originally for epidemiology. Here, based on my understanding of them, are Hill's criteria written in a way more applicable to general science:

(1) Temporality: Cause must occur before the effect.
(2) Strength: The larger the association is, the more likely that it is causal.
(3) Consistency: If the association is observed repeatedly under the same or various conditions (e.g., different persons, places, circumstances, and time), it is more likely causation.
(4) Specificity: Causation is more likely under specific conditions (e.g., a drug will more likely improve the health of certain diseased population than that of the general patient population), especially when the outcome has no other likely explanation.
(5) Response Curve: Given the causation, check if the cause–effect relation follows the expected response trend. Hill originally termed this criterion "Biological gradient" or dose-response as it is used for epidemiology, e.g., an increase in smoking will increase the chance of lung cancer.
(6) Plausibility: A plausible mechanism between cause and effect can be a support for causation.
(7) Coherence: A causal conclusion should not seriously conflict with present substantive knowledge.
(8) Analogy: Other established cause–effect relationships may be used in analogies to support this causal relationship.
(9) Experiment: A randomized experiment may sometimes be necessary, since this is a better way to find and confirm causation than an observational study.

How are these criteria to be used? Hill (1965) stated: "None of my nine viewpoints can bring indisputable evidence for or against the cause-and-effect hypothesis." Researchers mistakenly take the criteria as indisputable evidence and forget the word *likely* in his statement (the one prior to the list), thus raising many criticisms by others who point out counterexamples.

In my opinion, Hill's list of criteria can be used as a quick screening tool to sort the priorities of potential causal effects when time and resources are limited. They are more applicable to early exploratory studies than to late confirmatory studies. Hill's criteria can also be used as "common knowledge" to cross-check the conclusions from small quantitative studies.

Exploratory research is research into a new unknown territory. It is used when we want to investigate something interesting but know little about it. Confirmatory research is often a hypothesis-based study to confirm findings from early exploratory studies. A confirmatory study is often larger and more costly, and requires a longer time to conduct than an exploratory study. Qualitative research is usually exploratory in nature, while quantitative analysis can be used in both exploratory and confirmatory research.

In determining whether a quantitative or qualitative study should be used for solving the problem at hand, the decision depends on several factors: the goal of research (is a quantitative answer necessary or is a qualitative answer sufficient to the problem), the subject field (e.g., quantitative analyses are used more often in natural science than in social science, more often in physics than in psychology), the stage of the research or how much one knows about the problems (an earlier study is often qualitative in nature, while a quantitative study is more often used for a later confirmatory study), and time and cost. The different characteristics of the two approaches are outlined in the diagram.

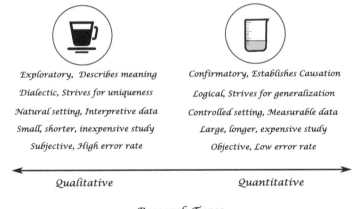

Exploratory, Describes meaning	Confirmatory, Establishes Causation
Dialectic, Strives for uniqueness	Logical, Strives for generalization
Natural setting, Interpretive data	Controlled setting, Measurable data
Small, shorter, inexpensive study	Large, longer, expensive study
Subjective, High error rate	Objective, Low error rate
Qualitative	Quantitative

Research Types

Heated debates have often occurred regarding which method should be used, qualitative or quantitative, especially in the field of sociology in the 1980s and 1990s. In my view, the two approaches are different lines of sight toward the same point. Ideally, if resources and time allowed, we should conduct both qualitative and quantitative research types (what is often called pluralistic research), since they provide different perspectives and usually complement each other. A quantitative study without earlier qualitative support may turn out to be much more costly, inefficient, or completely failed. A qualitative study without a follow-up quantitative study may display a lack of conclusive results, a lack of predictability, and limited usefulness. Sometimes, a pluralistic research that combines the two approaches may be a better choice.

Qualitative approaches to research in psychology and the social sciences are increasingly used (Frost et al., 2010), but the quantitative method still dominates research in most natural scientific fields. Statistics plays an important role in research, specially in quantitative analyses. Just as Professor David Hand (2008) pointed out: "Modern statistics... is an exciting subject which uses deep theory and powerful software tools to shed light on nearly every aspect of our lives—from astronomy to medical research, and from sociology to party politics and big business."

Chapter 2

Formal Reasoning

2.1 Mathematics as Science

Mathematics is the science that deals with the logic of shape, quantity, and their changes or dynamics. Mathematics may be studied in its own right (*pure mathematics*) or as it is applied to other disciplines (*applied mathematics*, e.g., statistics). The so-called *experimental mathematics*, in which numerical computation is used to investigate mathematical objects and identify properties and patterns, can do what pure and applied mathematics do. There is a range of views as to the exact scope and definition of mathematics.

Gauss referred to mathematics as the Queen of the Sciences. The specialization restricting the meaning of science to natural science follows the rise of Baconian science, which contrasts the Aristotelean method of inquiring from first principles (axioms or postulates). The role of empirical experimentation and observation is negligible in mathematics compared to natural sciences such as psychology, biology, or physics. But, as Einstein has put it, "How can it be that mathematics, being after all a product of human thought which is independent of experience, is so admirably appropriate to the objects of reality?" Einstein concluded, somewhat metaphysically perhaps, that "as far as the laws of mathematics refer to reality, they are not certain; and as far as they are certain, they do not refer to reality."

Many philosophers believe that mathematics is not experimentally falsifiable and is, thus, not a science according to the definition of Karl Popper (Section 1.4). However, Popper (2002) himself concluded that "most mathematical theories are, like those of physics and biology, hypothetico-deductive: pure mathematics therefore turns out to be much closer to the natural sciences whose hypotheses are conjectures, than it seemed even recently."

In many cases, mathematics shares much in common with many fields in the physical sciences, notably the exploration of the logical consequences of assumptions. Intuition and experimentation also play a role in the formulation of conjectures in both mathematics and other sciences. Experimental mathematics continues to grow in importance within mathematics, and computation and simulation are playing an increasing role in both the sciences and mathematics, weakening the objection that mathematics does not use the scientific method (wikipedians, 2013).

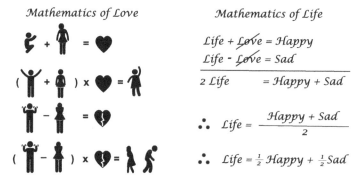

Mathematics of Love

Mathematics of Life

$$Life + Love = Happy$$
$$Life - Love = Sad$$
$$2\,Life \qquad = Happy + Sad$$

$$\therefore\ Life = \frac{Happy + Sad}{2}$$

$$\therefore\ Life = \tfrac{1}{2}\,Happy + \tfrac{1}{2}\,Sad$$

Math Loves Life

In his 1960 article, "The Unreasonable Effectiveness of Mathematics in the Natural Sciences," the physicist Eugene Wigner (1960) stated that the mathematical formulation of the physicist's often crude experience leads in an uncanny number of cases to an amazingly accurate description of a large class of phenomena. As an example, he invokes the law of gravitation, which was originally used to model freely falling bodies on the surface of the earth, and was extended, on the basis of very scanty observations, to describe the motion of the planets, where it has proved accurate beyond all reasonable expectations. Another example is Maxwell's equations, which, derived to model the elementary electrical and magnetic phenomena, also describe radio waves.

Wigner's original paper has provoked and inspired many responses across a wide range of disciplines, including mathematics, computer science, molecular biology, data mining, physics, and economics. Richard Hamming, an applied mathematician and a founder of computer science, states that the belief that science is experimentally grounded is only partially true. Rather, our intellectual apparatus is such that much of what we see comes from the glasses we put on. Hamming believes scientific discoveries arise from the mathematical tools employed and not from the intrinsic properties of physical reality. He gives four examples of nontrivial physical phenomena: (1) Hamming suspects that Galileo discovered the law of falling bodies not by experimenting, but by the simple thought experiment described in Section 2.6. (2) The inverse square law of universal gravitation necessarily follows from the law of conservation of energy. (3) The inequality at the heart of the uncertainty principle of quantum mechanics follows from the properties of Fourier integrals and from assuming time invariance. (4) Albert Einstein's pioneering work on special relativity was largely a thought experiment in its approach. He knew from the outset what the theory should look like and explored candidate theories with mathematical tools, not actual experiments.

It is a miracle that the human mind can string a thousand arguments together in mathematics without getting itself into contradictions. I completely agree with Wigner: whether humans checking the results of humans can provide an objective basis for observation of the known (to humans) universe is an interesting question. In Section 2.9, I will use mathematics' Pigeonhole Principle to suggest that it is impossible in theory for a human to prove anything that requires a sufficiently long step (longer than the brain can store). An obvious example is that when the theorem itself requires such a long string simply to express itself that exceeds the capability of a human brain.

Gödel proved that a mathematical system constructed on axioms and Aristotle logic (deductive reasoning) is incomplete. Thus, most mathematicians believe that an axiomatically based mathematical system can be useful even if incomplete, but it must be consistent. However, before we can identify all Gödel's unprovable statements or theorems in a mathematical system, how confident can we be to say that the system is consistent? Furthermore, we pursue mathematics as public knowledge for its maximum consistency, but 100% consistency may not be achievable due to some neurological deficiency in our brain from time to time. Maybe we can approach perfect consistency in this realm only over time, just as we approach truth through scientific discoveries over time.

It is so hard for me to reconcile between the two: (1) the human mind's capacity to divine mathematics—not a science whose truth involves uncertainty and (2) the study of brains being considered a science. In analogy, what we are saying here is that the size of a table (mathematics) measured with an inaccurate ruler (neuroscience) can always be exact.

2.2   Induction

It was Francis Bacon who first saw clearly that modern scientific method embodies a logic fundamentally different from that of Aristotle—induction. Bacon extols the method of careful observation and experimentation, from which inductions are made.

Induction, or *inductive reasoning*, is the process by which generalizations are made based on individual instances. The philosophical definition of inductive reasoning is much more nuanced than simple progression from particular or individual instances to broader generalizations. Rather, the premises of an inductive logical argument indicates some degree of support (inductive probability) for the conclusion but does not entail it; that is, it suggests truth but does not ensure it. Inductive reasoning consists of inferring general principles or rules from specific facts. In this section, we discuss only some intuitive inductive reasoning. More intricate probabilistic reasoning will be discussed in Chapter 4 under the heading of scientific inference.

Inductive Reasoning

Inductive reasoning is probabilistic; it states only that, given the premises, the conclusion is probable. The simplest induction is the so-called statistical syllogism.

A statistical syllogism is a form of inductive reasoning. A simple example would be

> Most humans are right-handed.
> Monica is a human.
> ───────────────────────────────
> Therefore, Monica is probably right-handed.

The statistical syllogism can be generally stated as

> A proportion Q of population P has attribute B.
> An individual X is a member of P.
> ───────────────────────────────
> Therefore, the probability that X has B is Q.

A meta-analysis of 144 studies including more than 1.7 million total participants found that about 90% of people are right-handed; thus we can make the following induction:

> 90% of humans are right-handed.
> Monica is a human.
> ───────────────────────────────
> Therefore, the probability that Monica is right-handed is 90%.

The same meta-analysis shows a gender correlation with handedness. According to the 2008 study, men have 1.23 times the odds of being left-handed, compared to women's chances. Therefore, we can make another induction about Monica's dominant hand.

> 92% of women are right-handed.
> Monica is a woman.
> ───────────────────────────────
> Therefore, the probability that Monica is right-handed is 92%.

"Induction is a critical tool for scientific discovery. We all believe that we have knowledge of facts extending far beyond those we directly perceive. The scope of our senses is severely limited in space and time; our immediate perceptual knowledge does not reach to events that happened before we were born to events that are happening now in certain other places or any future events" (Salmon, 1967).

Statistical syllogism is a pretty naive approach. For instance, even if a physician has cured the only cancer patient he has seen and so can boast a 100% cure rate, we wouldn't conclude that he will definitely cure his next cancer patient. How much the premises support the conclusion depends at least upon (a) the number of subjects in the sample (the sample size), (b) the population size, and (c) the degree to which the sample represents the population.

Abduction is the process of finding the best explanation from a set of observations, i.e., inferring cause from effect. In other words, abductive reasoning works in the reverse direction from deductive reasoning in which effect is inferred from cause. For instance,

> If and only if A is true, then B becomes truth.
> B is true.
> _____
> Therefore, A must be true.

Abduction often involves probability; thus, we have probabilistic abduction or plausible reasoning, which can take many different forms. For instance,

> If A is true, then B becomes more plausible.
> B is true.
> _____
> Therefore, A becomes more plausible.

A second formulation would be

> If A is true, then B becomes more plausible.
> A becomes more plausible.
> _____
> Therefore, B becomes less plausible.

Analogical reasoning is another kind of inductive reasoning which can be framed as

> P and Q are similar in respect to properties A, B, and C.
> Object P has been observed to further have property X.
> _____
> Therefore, Q probably has property X also.

Analogical reasoning is used very frequently in our daily lives and in the process of scientific discovery. A refined approach is case-based reasoning—a rapidly developing field in computer science.

Inductive inference is reasoning from the observed behavior of objects to their behavior when unobserved. Therefore, scholars such as David Hume (2011) criticized induction beginning with a simple question: How do we acquire knowledge of the unobserved? In Popper's view (2002), science is a deductive process in which scientists formulate hypotheses and theories that they test by deriving particular observable consequences.

2.3 Deduction

Deduction, or *deductive reasoning*, is reasoning whose conclusions are intended to necessarily follow from a set of premises. Deductive reasoning "merely" reveals the implications of propositions, laws, or general principles, so that, as some philosophers claim, it does not add up to truth. Unlike inductive reasoning which examines many pieces of specific information to impute a general principle, deductive reasoning applies general principles to reach specific conclusions.

Deductive Reasoning

An example of a deductive argument would be

All men are mortal.
John is a man.
Therefore, John is mortal.

The law of detachment and the law of syllogism are perhaps the simplest and most commonly used rules in making deductive arguments. The law of detachment (*modus ponens*) is the first form of deductive reasoning. A single conditional statement is made, and a hypothesis (P) is stated. The conclusion (Q) is then deduced from the statement and the hypothesis. The most basic form of the law of detachment can be written as

If P then Q (conditional statement).
P (hypothesis stated).
Therefore, Q (conclusion deduced).

Conversely, if Q is not true, then P is not true.

A simple application of the law of detachment would be

> If Joe is sick, he will not attend the class.
> Joe is sick today.
> ___
> Therefore, Joe will not attend the class today.

Conversely, if Joe attends the class, then Joe is not sick.

The law of syllogism takes two conditional statements and forms a conclusion by combining the hypothesis of one statement with the conclusion of another. Here is the general form, with the true premise P:

> If P then Q.
> If Q then R.
> ___
> Therefore, if P then R.

The following is an example:

> If Joe is sick, then he will be absent from school.
> If Joe is absent, then he will miss his class work.
> ___
> Therefore, if Joe is sick, then he will miss his class work.

We deduced the final statement by combining the hypothesis of the first statement with the conclusion of the second statement.

A more interesting example is about the number of walks required in a network. A famous problem concerning the seven bridges in Königsberg, Germany, was whether it was possible to take a walk through the town in such a way as to cross over every bridge once and only once.

Euler's approach was to regard the spots of land (there are 4 of them) as points to be visited and the bridges as paths between those points. The mathematical essentials of the map of Königsberg can then be reduced to the following diagram, which is an example of what is called a graph. For each of the vertices of a graph, the order of the vertex is the number of edges at that vertex. A walk is a sequence of links concerning a sequence of nodes (vertices) in a graph or network. During any walk in the graph, the number of times one enters a nonterminal node equals the number of times one leaves it. Therefore, one walk can at most remove two odd-degree nodes (terminals). Knowing that, can we determine the minimum number of walks required to cover all the paths in the figure? Here is the deductive reasoning employed to find out the answer.

> One walk will eliminate two odd-degree nodes.
> There are 4 odd-degree nodes.
> ___
> Therefore, a minimum of two walks is required to cover all the paths.

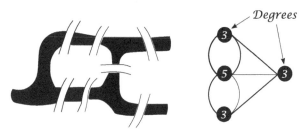

The Seven Bridge Problem

In a real world problem, the applications of deduction are often more complicated and full of traps. The following are two oversimplified examples.

When a scientist claimed to have invented an omnipotent solvent that could dissolve everything in the world, Edison asked, "What container will you use to store it if it dissolves everything?" Did Edison make a good logical argument?

At first thought, we may think that Edison made a very clever logical argument and proved that there is no such omnipotent solvent. However, anyone who has a little knowledge of chemistry would have a different answer, because there are many possible ways to store the solvent before we use it, for instance, storing at low temperature or in the dark to inhibit the chemical reactions.

Here is another interesting story. A biologist was conducting an experiment to study the behaviors of fleas. He was holding a flea in his hand and commanded it to jump; the flea indeed jumped. Then the biologist cut the flea's legs off and commanded it to jump again; the flea didn't jump. Therefore, he concluded that a flea becomes deaf after it loses its legs. We know that there are many possible logical interpretations of a phenomenon, but there is only one "correct" explanation.

2.4   Logic Computation

Since, for complex problems, natural language is difficult to use for reasoning, propositional calculus is used in mathematics and reasoning with the aid of a computer. A *proposition* is the natural reflection of a fact, expressed as a sentence in a natural or artificial language. Proposition calculus makes it possible for computing logic outcome with computer. Every proposition is considered to be true (1) or false (0).

Truth Table of Elementary Propositions

A	B	Conjunction $A \wedge B$	Disjunction $A \vee B$	Implication $A \longrightarrow B$	Equivalence $A \longleftrightarrow B$	Negation $\neg A$
1	1	1	1	1	1	0
1	0	0	1	0	0	0
0	1	0	1	1	0	1
0	0	0	0	1	1	1

In mathematical logic, *propositional calculus* is the study of statements, also called propositions, and compositions of statements. The statements are combined using *propositional connectives* such as *and* ($\wedge$), *or* ($\vee$), *not* ($\neg$), *implies* ($\longrightarrow$), and *if and only if* ($\longleftrightarrow$). Propositions or statements are denoted by $\{A, B, C, \ldots\}$. The precedence of logical operators (connectives) are parentheses, $\neg$, $\wedge$, $\vee$, $\longrightarrow$, $\longleftrightarrow$. Sometimes, we omit $\wedge$ for simplicity. The *conjunction, disjunction, implication, equivalence,* and *negation* in relation to their component elements or events are self-explanatory as summarized in the table above. For example, the conjunction means that both A and B are true. If one of them is not true, the conjunction is not true. The disjunction means that either A or B is true. If both A and B are false, then the disjunction is false; otherwise it is true. The implication $A \longrightarrow B$ is true unless A is true and B is false. It is obvious that the conjunction implies the disjunction, but the converse is not true. Keep in mind that when A is false, then the implication $A \longrightarrow B$ is true, regardless of B. The implication of this is that *truth* can include what we cannot prove to be wrong and is not just limited to what can be proved to be correct. For instance, suppose you claim that you would have won the lottery if you had bought a lottery ticket. This claim cannot be proved wrong (also cannot be proved correct), since you didn't buy the ticket; therefore, the claim is valid or true.

After we code 1 for truth and 0 for false, we can easily convert the propositions into arithmetic expressions as shown in the following diagram. The pairs of expressions are equivalent for any logic values (true and false) of A and B.

With propositional calculus, we will see how easy it is to convert a formal reasoning problem into an arithmetic problem so that it can be handled by a computer, as it has been done successfully in the proof of the famous four-color problem.

The *four-color theorem*, motivated from the coloring of political maps of countries, can be stated intuitively: In a map with contiguous regions, the regions can be colored using at most four colors so that no two adjacent regions have the same color. For simpler maps, only three colors are necessary. The four-color theorem was proven in 1976 by Kenneth Appel and Wolfgang Haken using computer algorithms they developed. It was the first major theorem proved using a computer.

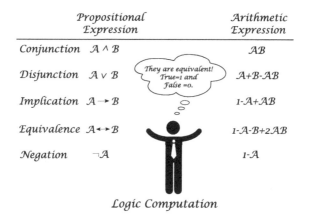

	Propositional Expression	Arithmetic Expression
Conjunction	$A \wedge B$	AB
Disjunction	$A \vee B$	$A + B - AB$
Implication	$A \rightarrow B$	$1 - A + AB$
Equivalence	$A \leftrightarrow B$	$1 - A - B + 2AB$
Negation	$\neg A$	$1 - A$

Logic Computation

Many laws that we use consciously or subconsciously in our daily lives, and in scientific research, can be expressed in term of axioms or mathematical tautologies. A *tautology* is a statement which always is true. Some of the most commonly used tautologies are

- *Law of excluded middle*: $A \vee \neg A$. It says, e.g., that one of these two statements must be true: "You have the book." or "You do not have the book."
- *Law of detachment (modus ponens)*: $((A \longrightarrow B) \wedge A) \longrightarrow B$. It says, e.g., that if you have the book, you must have read it. You have the book; therefore, you have read it.
- *Reductio ad absurdum (proof by contradiction)*: $(\neg A \longrightarrow A) \longrightarrow A$. It says that if the negation of A implies A, then A is true. The notion behind this law is the consistency axiom, which asserts that A and $\neg A$ cannot be simultaneously true.

These elementary laws can be easily verified using the corresponding arithmetic expressions. For instance, the excluded middle law, $A \vee \neg A \equiv 1$, can be proved as follows: Starting with the disjunction $A \vee B = A + B - AB$, we use the conversion $\neg A = 1 - A$ to replace B on the left side of the disjunction with $\neg A$ and B on the right side with $1 - A$, leading to

$$A \vee \neg A = A + (1 - A) - A(1 - A) = 1 - A(1 - A) = 1.$$

Similarly, the law of detachment is verified in this way:

$$((A \longrightarrow B) \wedge A) \longrightarrow B = 1 - (1 - A + AB)A + (1 - A + AB)AB = 1.$$

For developing the logical foundations of mathematics, we need a logic that has a stronger expressive power than propositional calculus. *Predicate calculus* is the language in which most mathematical reasoning takes place. *Predicates* are variable statements that may be true or false, depending on the values of their variables. Predicate calculus uses the universal quantifier ∀, the existential quantifier ∃, the symbol ∈ for "is an element of" or "belongs to," predicates $P(x)$, $Q(x, y)$, variables $x, y, ...$, and assumes a universe U from which the variables are taken.

2.5 Mathematical Induction

Mathematical induction is a widely used tool of deductive (not inductive) reasoning. Mathematical induction is a method of mathematical proof typically used to establish that a given statement is true for all (or some) natural numbers. It involves the following two steps:

(1) The basis: Providing the statement to be proved is a function of the natural number n, prove the statement is true for a natural number n_0.
(2) The inductive step: Assuming the statement is true for any natural number $n \geq n_0$, prove the statement is true for $n + 1$.

Taking arithmetic progression as an example, prove the following identity holds for any positive integer n:

$$1 + 3 + \cdots + 2n - 1 = n^2.$$

The basis: It is obvious that for $n = 1$ the above equation holds.
The inductive step: If the above equation holds for n, we prove that it holds for $n + 1$. In fact,

$$(1 + 3 + \cdots + 2n - 1) + [2(n + 1) - 1] = n^2 + (2n + 1) = (n + 1)^2.$$

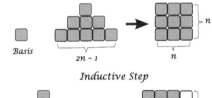

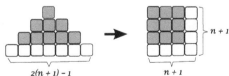

Mathematical Induction

Another form of mathematical induction is the so-called complete induction, in which the second step is modified to assume that the statement holds not only for n, but also for all m satisfying $n_0 \leq m \leq n$.

Despite the simplicity of mathematical induction, mistakes can often happen. The following are two examples of false mathematical induction arguments by an anonymous author.

The first example is about the proof of the statement "everything in the world has the same color."

The basis: The statement holds for $n = 1$ object because any object has the same color as itself.

The induction step: Suppose the statement holds when $n = k$, i.e., any k things in the world are of the same color. Now consider the case $n = k + 1$. Take $k + 1$ things in the world and line them up in a row. By the induction hypothesis, the first k things have the same color. By the induction hypothesis again, the last k things have the same color. Therefore, the $k + 1$ things are of the same color.

By the principle of mathematical induction, the statement holds for all positive integers n. That is, everything in the world has the same color!

The second example is about the proof of superpower.

There are two positive integers differing by 1 (e.g., 3 and 4) written on two different papers for two gentlemen to choose. The rule of the game requires that if anyone knows, without seeing the other number, that his number is the smaller of the two, then he should raise his hand. In other words, if two people are given positive integers n and $n + 1$, the one having the integer n will raise his hand. We are going to prove that the one given n will raise his hand by mathematical induction—he has superpower knowing the number he didn't see.

The basis: This is true when $n = 1$, because 1 is the smallest positive integer, so the one given the number 1 must know that his number is the smaller of the two.

Suppose the statement holds when $n = k$. Consider the case $n = k + 1$. Then the two people are given $k + 1$ and $k + 2$. The one with $k + 1$ will think, "The other number is either k or $k + 2$. If it is k, then by the induction hypothesis, the other person should have raised his hand. But he has not, so the other number must be $k + 2$." Consequently, the one with $k + 1$ will raise his hand.

By the principle of mathematical induction, humans possess this superpower.

Mathematical induction has been generalized to a method called structural induction in mathematical logic, graph theory, and computer science, where recursion is applied to structures beyond the natural numbers, such as structural trees. Structural recursion is a recursion method bearing the same relationship to structural induction as ordinary recursion bears to ordinary mathematical induction.

Suppose we wish to prove some proposition $P(x)$, where x is any instance of some sort of recursively defined structure such as a list or tree. The structural induction proof is a proof that the proposition holds for all the minimal structures, and that if it holds for the immediate substructures of a certain structure S, then it must hold for S also.

2.6 Thought Experiments

Einstein (1933, p. 144) said: "Pure logical thinking cannot yield us any knowledge of the empirical world; all knowledge starts from experience and ends in it. Propositions arrived at by purely logical means are completely empty as regards reality."

This must be understood in context. In general, knowledge can be gained from both experience and logical reasoning (induction and/or deduction). More importantly any new piece of experience can become new knowledge only through some sort of logical reasoning, be it simple or complex. Any new knowledge obtained through logical reasoning (induction and/or deduction) has to be based on some sort of old or new experiences. Let's see a simple example using the omnipotence paradox to show how new knowledge can be obtained through deductive reasoning.

The *omnipotence paradox* can be stated as: "Can an omnipotent being create a rock so heavy that it cannot lift it?" This question generates a dilemma. The being can either create a stone which it cannot lift, or it cannot create a stone which it cannot lift. If the being can create a stone that it cannot lift, then it seems that it ceases to be omnipotent. If the being cannot create a stone which it cannot lift, then it seems it is already not omnipotent. The deductive reasoning technique we used here to disprove the statement "God is omnipotent" is *proof by contradiction* discussed in Section 2.10.

What new knowledge have we gained from the deduction? We have learned that we can keep only one of the two: an omnipotent God or deductive reasoning.

The omnipotence paradox can be considered as a special type of "experiment," called a thought experiment. A thought experiment typically includes a virtual experiment setting, a sequence of deductive reasoning (induction is rarely used), and a conclusion. Given the structure of the experiment, it may or may not be possible to actually perform it. The common goal of a thought experiment is to explore the potential consequences of the principle in question. Thought experiments have been used in philosophy, physics, cognitive psychology, political science, economics, social psychology, law, marketing, and epidemiology. They can be used to (1) challenge a prevailing status quo or confirm a prevailing theory, (2) establish a new theory, (3) extrapolate beyond the boundaries of already established fact, (4) predict and forecast the (otherwise) indefinite and unknowable future, and (5) explain the past.

Galileo's Leaning Tower of Pisa experiment was used for rebuttal of Aristotelian gravity. Galileo showed that all bodies fall at the same speed with a brilliant thought experiment that started by destroying the then reigning Aristotelian account (James R. Brown, 1991). The latter holds that heavy bodies fall faster than light ones ($H > L$). But consider this: in which a heavy cannon ball (H) and a light musket ball (L) are attached together to form a compound object ($H + L$); the last must fall faster than the cannon ball alone. Yet the compound object must also fall slower, since the light part will act as a drag on the heavy part. Now we have a contradiction: $H + L > H$ and $H > H + L$. That is the end of Aristotle's theory. But there is a bonus, because the right account is now obvious: they all fall at the same speed ($H = L = H + L$).

Thought Experiment of Falling Balls

The second example is a problem in physics. We know that matter can transit between phases (gas, liquid, or solid) as temperature changes. For instance, when temperature rises ice becomes water, and water becomes vapor. Interestingly, we can deduce, through proof by contradiction, that there must be a temperature at which both the gas and liquid states of a substance coexist. We know that if a substance can exist at two temperatures, then it can exist at any temperature in between (the continuity principle of temperature). Now assume the lowest temperature for the liquid state of the substance is C_l and the highest temperature for the solid state of the substance is C_h. If C_l and C_h are different, then there is no such state of matter at temperature $C = (C_l + C_h)/2$. That is contradictory to the continuity principle of temperature. Therefore, C_l and C_h must be the same.

2.7   Fitch's Knowability Paradox

According to William Byers (2007, p. 343), "Knowing the truth" is a single unity—both an object and an event, objective and subjective. Knowing and truth are not two; they are different perspectives on the same reality. There is no truth without knowing and no knowing without truth. In other words, the truth is not the truth unless it is known. Nevertheless "truth" and "knowing" are not identical. They form an ambiguous pair—one reality with two frames of reference.

Fitch's Knowability Paradox is a logical result suggesting that, necessarily, if all truths are knowable in principle then all truths are in fact known. The contrapositive of the result says, necessarily, if in fact there is an unknown truth, then there is a truth that couldn't possibly be known. Fitch's Paradox has been used to argue against versions of anti-realism committed to the thesis that all truths are knowable. It has also been used to draw more general lessons about the limits of human knowledge.

Suppose p is a statement which is an unknown truth; it should be possible to know that "p is an unknown truth." But this isn't possible; as soon as we know "p is an unknown truth," we know p, and thus p is no longer an unknown truth and "p is an unknown truth" becomes a falsity. The statement "p is an unknown truth" cannot be both verified and true at the same time.

Fitch's Knowability Paradox

The proof of Fitch's Paradox can be formalized with modal logic (see Chang, 2012). We now have to face a choice: (1) all truths are known already (equivalently, all truths are knowable in principle), or (2) not all truths are knowable.

Personally, I choose (2). In fact, using deductive reasoning, Kurt Gödel proved (the Incompleteness Theorem) that there is some theorem which either is not provable or we will never know if it is true or not.

2.8 Incompleteness Theorem

Kurt Gödel's beautiful *Incompleteness Theorem* completely surprised the mathematical world at the time it was published. The theorem is definitely an archetypal example of how new knowledge can be found purely using deductive reasoning without new observations.

Gödel used the liar paradox (in mathematical form) to prove one of the most shocking results in mathematics. The *liar paradox* is the statement "this sentence is false." It is paradoxical because: if "this sentence is false" is true, then the sentence is actually false. Conversely, if "this sentence is false" is false, then it means that the sentence is true.

Gödel proved in 1931 that any formal system that is complicated enough to deal with arithmetic, that is, contains the counting numbers 0, 1, 2, 3, 4, ... and their properties under addition and multiplication, must either be inconsistent or incomplete. Gödel's incompleteness theorem is important both in mathematical logic and in the philosophy of mathematics. It is one of the great intellectual accomplishments of the twentieth century. Its implications are so far reaching that is difficult to overestimate them.

Liar's Paradox

Gödel's incompleteness proof is very clever. He first invents a brilliant way of coding the elements (symbols, statements, propositions, theorems, formulas, functions, etc.) of the formal system into numbers (called Gödel numbers) and arithmetical operations. Conversely, each such integer can be decoded back into the original piece of formal mathematics. Thus, the proof or disproof of a statement is equivalent to the proof of the existence or absence of the corresponding Gödel number. Next, what Gödel did was to construct a self-referencing paradoxical statement that says of itself, "I'm unprovable!" in arithmetic.

Let $G(F)$ denote the Gödel number of the formula (statement, theorem, etc.) F. For every number n and every formula $F(x)$, where x is a variable, we define $R(n, G(F))$ as the statement "n is not the Gödel number of a proof of $F(G(F))$." Here, $F(G(F))$ can be understood as F with its own Gödel number as its argument.

If $R(n, G(F))$ holds for all natural numbers n, then there is no proof of $F(G(F))$. We now define a statement P by the *Gödel-Liar equation:*

$$P(G(P)) = R(n, G(P)) \text{ for all } n.$$

The *Gödel-Liar equation* concerning the number $G(P)$ is the Liar Paradox: $P(G(P))$ says: "I am not provable." If this Gödel-Liar equation is provable, its left-hand side says $P(G(P))$ is valid, but its right-hand side says "there is no proof of $P(G(P))$" because no *Gödel number exists for* $G(P)$! This violates the consistency axiom of the formal theory. Therefore, the Gödel-Liar equation is unprovable for some formula (function) F that has Gödel numbers as its variables. Thus, the formal system is incomplete.

In conclusion, Gödel's incompleteness theorem proves that a formal system is either inconsistent or incomplete. However, we seem to be more tolerant of incompleteness than inconsistency, and that may be why we have Gödel's Incompleteness Theorem instead of Gödel's Inconsistency Theorem. A consistency proof for any sufficiently complex system can be carried out only by means of modes of inference that are not formalized in the system itself. As Gödel said, "Either mathematics is too big for the human mind or the human mind is more than a machine."

However, Gödel's theorems apply only to effectively generated (that is, recursively enumerable) theories. If all true statements about natural numbers are taken as axioms for a theory, then this theory is a consistent, complete extension of Peano arithmetic (called true arithmetic) for which none of Gödel's theorems hold, because this theory is not recursively enumerable. Dan E. Willard (2001) has studied many weak systems of arithmetic which do not satisfy the hypotheses of Gödel's equally profound Second Incompleteness Theorem, and which are consistent and capable of proving their own consistency.

2.9 Pigeonhole Principle

The *pigeonhole principle* can be simply stated as: If one wishes to put n pigeons into m holes ($0 < m < n$), there will be a hole that contains at least 2 pigeons. The pigeonhole principle is also commonly called *Dirichlet's drawer principle,* in honor of Johann Dirichlet who formalized it in 1834 as "the drawer principle."

The pigeonhole principle has many applications. Simple examples would be (1) Among any 13 people there are at least two having the same birth month.

(2) at any cocktail party with two or more people, there must be at least two who have the same number of friends (assuming friendship is a mutual thing). (3) If some number of people shake hands with one another, there is always a pair of people who will shake hands with the same number of people.

Here is another example: When placing 5 pigeons in a 2ft×2ft×1ft cage, there will be at least two pigeons, who, when standing, are at most $\sqrt{3}$ ft apart from each other.

Proof: Divide the cage into four 1ft×1ft×1ft cubical rooms. Based on the pigeonhole principle, there are at least 2 pigeons in one of the rooms. It follows that the distance between the two birds is at most $\sqrt{3}$ ft, when they stand in opposite corners of the room.

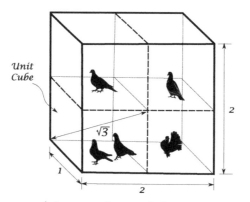

Application of Pigeonhole Principle

The pigeonhole principle is useful in computer science. For instance, the principle proves that any general-purpose lossless compression algorithm that makes at least one input file smaller will make some other input file larger. Otherwise, the two files would be compressed to the same smaller file and restoring them would be ambiguous.

A generalized version of the pigeonhole principle states that, if n discrete objects are to be allocated to m containers, then at least one container must hold no fewer objects than the smallest integer not less than n/m and no more objects than the largest integer less than or equal to n/m.

A probabilistic generalization of the pigeonhole principle states that if n pigeons are randomly put into $k > n$ pigeonholes, then the probability that at least one pigeonhole will hold more than one pigeon is

$$1 - \frac{k(k-1)(k-2)...(k-n+1)}{k^n}.$$

For example, if $n = 2$ pigeons are randomly assigned to $k = 4$ pigeonholes, there is a 25% chance that at least one pigeonhole will hold more than one pigeon.

2.10 Proof by Contradiction

We have discussed proof by contradiction earlier: If the negation of A implies A, then A is true. An intelligent application of proof by contradiction is *Russell's Paradox*. Discovered by Bertrand Russell in 1901, and sometimes called the *barber paradox*, it presents an apparently plausible scenario that is logically impossible. Suppose there is a town with just one male barber who claims that he will shave every man in the town who never shaves himself, but will not shave any man who shaves himself. This sounds very logical until the following question is asked: Does the barber shave himself?

If the barber does not shave himself, he must abide by his rule and shave himself. However, if he does shave himself, then according to the rule he should not shave himself.

Russell's Paradox, restated in the context of the so-called "naïve" set theory, exposed a huge problem in the field of logic and changed the entire direction of twentieth-century mathematics.

There are two types of sets in the naïve set theory. For the first type of sets, the set is not a member of itself. For instance, the set of all nations is not a nation; the set of all men is not a man. For the second type of set, the set is a member of itself. For instance, the set of everything which is not a human is a member of itself. Now what about the set of all sets which are not members of themselves? Is it a member of itself or not? If it is, then it isn't, and if it isn't, then it is ... just like the barber who shaves himself, but mustn't, and therefore doesn't, and so must!

Russell's Paradox means that there is a contradiction at the heart of naïve set theory. That is, there is a statement S such that both itself and its negation (not S) are true, where the statement S is "the set of all sets which are not members of themselves contains itself." The problem is that once there is such a contradiction, one can prove anything being true using the rules of logical deduction! This is how it goes:

Suppose if S is true, then Q is true, where Q is any other statement. But "not S" is also true, so Q is true no matter what statement Q says. The paradox raises the frightening prospect that the whole of mathematics is based on shaky foundations, and that no proof can be trusted. The Zermelo–Fraenkel set theory with the axiom of choice was developed to escape from Russell's Paradox.

More surprisingly, reductio ad absurdum does not always work due to the pigeonhole principle and the fact that the brain is bounded. Just as Albert Einstein uses the limit of ultimate speed (the speed of light) to derive counterintuitive facts in relativity, here we can use a limit of brain states to derive counterintuitive conclusions.

It is reasonable that since memory (mind) is limited to, say, 100 billion, then one cannot completely differentiate the numbers 0 to 100 billion, on account of the pigeonhole principle. This can be proved by reductio ad absurdum. Certainly memory is also needed to map emotional and other states. For instance, a feeling for "the infinite" or conceiving of infinity is a state of mind. Therefore, we have a dilemma: reductio ad absurdum is used to prove the pigeonhole principle, which is used to disprove reductio ad absurdum.

Limitations of Our Brains

We should not confuse the limitation of brain states with the limitation of computer memory. If computer memory is full, we can erase it and put new content in. This is because our brain is larger than a computer, memory-wise. We know what is erased is different from what is put in computer memory later on. But brain state limits will not allow one to differentiate between too many things that are different, so the world will always appear to be limited, even if a person could live forever. For example, a brain state corresponding to a "sweet" feeling will still give the sweet feeling even if one is actually given a salty food, as long as this particular brain state is reached. Another example: if a person has only 8 brain states for numbers, he may see symbols 8 and 9 as the same because they correspond to the same brain state. Thus, 8 and 9 have no difference. To him, given the number ≥ 8, "Not 8" (i.e., 9) can imply "8;" thus, from reductio ad absurdum 9 is 8, even if he knows logically a computer can store more than 8 numbers, but he cannot see and cannot possibly see any number beyond 8.

Because memory and time are limited, we cannot think about thinking about thinking,..., indefinitely. This self-referential paradox suggests that Aristotle's logic is not conducive to thinking about thinking.

2.11 Dimensional Analysis

The *dimension* of a physical quantity is the combination of the basic physical dimensions (usually length, mass, time, electric current, temperature, amount of substance, and luminous intensity) which describe it. The dimension of a quantity is different from its particular units. For instance, speed has the dimension length per time and may be measured in meters per second, miles per hour, or other units. *Dimensional analysis* is based on the fact that a physical law must be independent of the units used to measure the physical variables.

In physics and engineering science, dimensional analysis is a tool used (1) to check the plausibility of derived equations and computations on the basis of dimensional homogeneity, (2) to simplify an equation that characterizes the relationships among variables based on the Buckingham π theorem, and (3) for out-of-scale modeling, to determine the properties and behaviors of the original via a prototype (often in a much smaller scale) in terms of the similarity law (not the similarity principle). Dimensional analysis relies on the Buckingham π theorem, while the same theorem is in turn a consequence of deductive reasoning, following from the simple notion that a physical law should be independent of the units chosen.

The most basic rule of dimensional analysis is dimensional homogeneity: Only commensurable quantities (quantities with the same dimensions) may be compared, equated, added, or subtracted. Scalar arguments of transcendental functions, such as exponential, trigonometric and logarithmic functions, or inhomogeneous polynomials, must be dimensionless quantities. An important consequence of dimensional homogeneity is that the form of a physical equation is independent of the size of the base units.

Dimensional analysis offers a method for reducing complex physical problems to the simplest form prior to obtaining a quantitative answer. The *Buckingham π theorem* is a key theorem in dimensional analysis. The theorem loosely states that if we have a physically meaningful equation involving $k+m$ variables and these variables are expressible in terms of k dimensionally independent fundamental quantities, then the original function is equivalent to an equation involving only a set of m dimensionless parameters constructed from the original variables, i.e.,

$$\pi_1 = f(\pi_2, ..., \pi_m).$$

Here we are not going to formally prove the theorem, but the main reason that the theorem is true is this: The equation characterizing the underlying physical relationship should be independent of the choice of the units and there are k fundamental units which can be chosen to determine the units for all the variables. This invariance of equation with respect to the k dimensional changes serves as a set of k constrains (equations) that can be used to eliminate k variables in the equation.

Buckingham's π theorem provides a method for computing sets of dimensionless parameters from the given variables, even if the form of the equation is still unknown. However, the choice of dimensionless parameters is not unique and is not made on the basis of any physical meaning.

The π theorem was first proved by Bertrand in 1878. Bertrand considered only special cases of problems from electrodynamics and heat conduction. Formal generalization of the π theorem for the case of an arbitrary number of quantities was completed by a number of scholars including Buckingham in 1914.

Out-of-scale modeling is a major application of dimensional analysis in engineering science that deals with the following question: If we want to learn something about the performance of a full-scale system by testing a geometrically similar small-scale system model (or vice versa, if the system of interest is inaccessibly small), under what conditions should we test the model, and how should we obtain full-scale performance from measurements at the small scale? At the heart of out-of-scale modeling is the similarity law. If the π theorem is followed in building a prototype, there is a direct relationship between a physical quantity (e.g., the forces acting on a spaceship) and that of a small-scale model of it (the prototype).

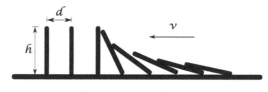

A Collapsing Row of Dominoes

Variable	Symbol	Dimension
Velocity of collapse	v	m/s
Separation	d	m
Height	h	m
Gravitational acceleration	g	m/s^2

Note: m = the mass dimension s = the time dimension

Suppose in planning an Olympic Game Opening, the artistic director wants to show a scene with a long row of huge dominoes falling. He needs to learn the speed and timing of the falling. However, running tests on a full scale is impossible because of cost and other limitations. He instead uses dimensional analysis to solve the problem easily. The following example is adapted from Szirtes (2007).

From mechanics, we know that velocity of collapse depends on the five variables: separation, height, thickness, gravitational acceleration, and density of the materials. Since we use the same material for the prototype and the full-scale dominos and because the thickness is negligible compared with the height h and the separation d, there are actually just one dependent and three independent variables needed as listed in the table. We can see from the table above that there are only two dimensions involved: length (m) and time (s). It follows that there are $4 - 2 = 2$ dimensionless variables required to establish the physical relationship. We choose h and g as the independent fundamental variables, from which we formulate two dimensionless quantities:

$$\pi_1 = \frac{v}{\sqrt{hg}} \text{ and } \pi_2 = \frac{d}{h}.$$

Per the π theorem, we can write the relationship between the dimensionless quantities as $\pi_1 = f(\pi_2)$ or

$$v = \sqrt{hg} \cdot f\left(\frac{d}{h}\right).$$

This means that, given the same constant $\frac{d}{h}$ in the prototype as in the original object, the velocity of collapse increases as the square root of the height h. It is interesting that v varies as the square root of g, i.e., the same dominoes (set up identically) would collapse on the Moon at a speed of only 40.3% of that on Earth. From this, we can see how spaceships for landing on the Moon can likewise be modeled on Earth.

Dimensional analysis is important and widely used in engineering and science. We will discuss this further from the perspective of biological scaling in Chapter 7.

Chapter 3

Experimentation

3.1 Overview of Experimentation

Although experiment without theory is blind, theory without experiment is empty. The experiment is the most commonly used tool in scientific research. The goal of an experimental design is to establish or confirm a causal connection between the independent and dependent variables with the minimum expenditure of resources. An experiment involves a number of interrelated sequential activities:

(1) Formulating research objectives into scientific hypotheses or statistical hypotheses. A statistical hypothesis can be tested for truth or falsehood using experimental data and sometimes with existing knowledge. An example of a hypothesis would be "weekly three-hour running exercise will reduce body weight."

(2) Specifying the target population if applicable, e.g., subjects with weight over 230kg for the above example.

(3) Determining the outcome (dependent) variables of interest and causal (independent) variables to be recorded. For the same example, we can choose body weight for the dependent variable and age, gender, race, number of timed exercising per week, and economic status as independent variables. In addition, there are so-called nuisance variables, which are the source of random errors.

(4) Determining the sample size (number of experimental units or subjects required), based on statistical method, and specifying the procedure, usually randomized, for assigning subjects to the experimental conditions. For the weight study, supposing that 200 subjects are required, we then can randomly (with equal probability) assign patients to the group of those who will do the exercise or the group who will not.

(5) Selecting facilities to conduct the experiment and carefully collect data. For the weight study, we may ask participants to come to selected gyms

to do the exercise and record their weight change daily or they can do the same exercise and record their weight by themselves.

(6) Planning and performing statistical analysis on the experimental data.

(7) Interpreting the results.

Experiments are usually performed in an environment that permits a high degree of control of confounding variables, which are the variables other than the main variable in which we are interested. Such environments rarely duplicate real-life situations, but in any case an experiment is still a useful way of obtaining knowledge. A key concept in experimentation is the factor isolation technique: the experimenter manipulates the experimental condition so that the outcomes come from two different conditions, with and without the main putative causes (independent variables), while other nuisance variables are balanced using the technique called randomization. The notion and usefulness of factor isolation is given by the so-called laws of factor isolation: given other things unchanged,

B implies W.	B implies W.
Not B implies not W.	Not B implies W.
Therefore, B is a cause of W.	Therefore, B is not a cause of W.

The *laws of factor isolation* can be illustrated in the following diagram:

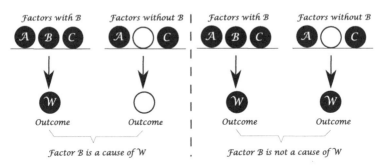

Laws of Factor Isolation

It is not always possible to make random assignments due to practical or ethical reasons. In such a case, it is necessary to use preexisting or naturally occurring groups such as subjects with a particular illness or subjects who have been sexually abused. This type of experiment is sometimes called a *Quasi-experiment*.

Surveys are another common type of research instrument, which relies on the technique of self-report to obtain information. The data are collected by means of an interview or a questionnaire. Although surveys cannot usually establish causality, they can explore, describe, classify, and establish correlation among variables.

In contrast to experiments, naturalistic observation involves observing individuals or events in their natural setting without using manipulative intervention.

An *observational study* can be classified into longitudinal and cross-sectional studies. In a *longitudinal study*, the same individuals are observed at two or more times. Since a longitudinal study usually involves studying the same individuals over a long period of time, "lose to follow-up" is a major challenge. Identifying the cause of the observed changes is another challenge since it is difficult to identify and control all nuisance variables over an extended period of time in the design or analysis. In a *cross-sectional study* two or more cohorts are observed at the same time. A cross-sectional study is applicable when the outcome can be observed in a relatively short time. Cross-sectional studies tend to be less expensive than longitudinal studies.

3.2   Experimentation in Life Science

Commonly used experiment types in the life sciences include in-vitro, in-vivo, and ex-vivo experiments; clinical trials; and epidemiological studies.

In vitro (in-tube) studies in experimental biology are those that are conducted using components of an organism that have been isolated from their usual biological surroundings in order to permit a more detailed or more convenient analysis than those that can be done with whole organisms. In contrast, *in-vivo* work is that which is conducted with living organisms in their normal, intact state, while *ex-vivo* studies are conducted on functional organs that have been removed from the intact organism.

The advantages of in-vitro work are that it is inexpensive to perform and it permits an enormous level of simplification of the system under study, so that the investigator can focus on a small number of components. The main disadvantage of in-vitro experimental studies is that it can sometimes be very challenging to extrapolate from the results of in-vitro work back to the biology of the intact organism. Furthermore, animals and human beings, despite some similarities between them, are different; therefore, the experimental results in animals cannot be directly extrapolated to humans without validation. That is why clinical trials are necessary for any drug candidate that is intended for human use.

Clinical trials are experiments on human beings to evaluate the efficacy and safety of health interventions (e.g., drugs, diagnostics, devices). They are the experiments immediately following the preclinical in-vivo experiments in drug development. Clinical trials are costly and having the highest ethical and scientific standards.

Clinical trials are usually conducted in steps or phases. Such a stepwise approach is due to cost and safety considerations. As data accumulate, if the test compound is demonstrated to be somewhat safe and/or efficacious, further experiments are granted with an increased number of patients.

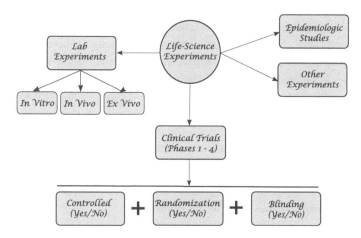

Experiments in Life Sciences

A *Phase-1 trial* is performed mainly to test the safety of the drug candidate. Researchers test an experimental drug or treatment in a small group (16–80) of people for the first time to evaluate its safety, determine a safe dosage range, and identify side effects. This kind of trial is typically also used to characterize the drug's pharmacokinetic and pharmacodynamic properties.

Phase-2 trials are used to further test for safety and efficacy in a larger population (100–300 subjects) and to find the optimal dose.

In a *Phase-3 trial*, the treatment is given to large groups of people (500–10,000) to confirm its effectiveness, monitor side effects, compare it to commonly used treatments, and collect information that will allow it to be used safely.

Phase-4 trials are called post-approval studies, which delineate additional information, including the treatment's risks, benefits, and optimal use.

There are three important concepts in clinical trials that generally fit most experiments: (1) randomization, (2) control, and (3) blinding. They are effective ways to control confounders and reduce bias in the design and conduct of a trial. The combination of these three makes up the gold standard for a scientific experiment.

In recent years, the cost for drug development has increased dramatically, but the success rate of new drug applications (NDAs) remains low. The pharmaceutical industry devotes great efforts in innovative approaches, especially on adaptive design. An *adaptive design* is a clinical trial design that allows adaptations or modifications to aspects of the trial after its initiation without undermining the validity and integrity of the trial (Chang 2007). The adaptation can be based on information internal or external to the trial.

The purposes of adaptive trials are to increase the probability of success, to reduce both the cost and the time to market, and to deliver the right drug to the right patient.

Unlike the kind of experimental study seen in clinical trials, *epidemiology* is the study of the patterns, causes, and effects of health and disease conditions in defined populations. It is often used to inform policy decisions by identifying the factors that are associated with diseases (risk factors), and the factors that may protect people or animals against disease (protective factors). Major areas of epidemiological study include *disease etiology, outbreak investigation, disease surveillance* and *screening*, and *biomonitoring*.

In general, however, epidemiological studies cannot prove causation. That is, it cannot be proved that a specific risk factor actually causes the disease being studied. Epidemiological evidence can only show that this risk factor is associated (correlated) with a higher incidence of disease in the population exposed to that risk factor. In a larger study, the higher the correlation the more certain the association, but a higher correlation cannot prove the causation because there are potential confounding factors (the factors that are unidentified but affect the disease outcomes) that are not controlled. To identify or verify the possible causes identified by epidemiological studies, a randomized, controlled study such as a clinical trial is usually required.

3.3 Control and Blinding

To explain the terms experimental *control* and *blinding*, we should start with the concept of the placebo effect.

A *placebo* is a substance or procedure that is objectively without specific activity for the condition being treated (Moerman and Jonas, 2002). In medical experiments, a placebo can be pills, creams, inhalants, and injections that do not involve any active ingredients. In studies of medical devices, ultrasound can act as a placebo. Sham surgery, sham electrodes implanted in the brain, and sham acupuncture can all be used as placebos.

The *placebo effect* is related to the perceptions and expectations of the patient. If the substance is viewed as helpful, it can heal, but if it is viewed as harmful, it can cause negative effects, which is known as the *nocebo effect*.

However, it is interesting that a study carried out by researchers in the Program in Placebo Studies at the Harvard Medical School discovered there is

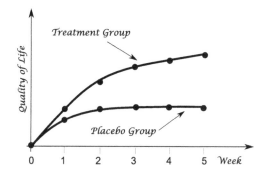

Treatment and Placebo Effects in an Anemia Study

an "*honest placebo*" effect: patients with irritable bowel syndrome in one study experienced a significant beneficial effect even though they were told the pills they were taking were placebos, as compared to a control group who received no pills (Kaptchuk et al., 2010). This is perfectly logical if we understand why many of us are afraid of walking in a graveyard in the dark even though we don't believe there are ghosts.

Placebo effects are generally more significant in subjective measurements, such as with patient-reported outcomes, than in an objective laboratory measurements such as a hemoglobin level.

There are controversies surrounding placebo use as medical treatment. A survey conducted in the United States of more than 10,000 physicians indicated that while 24% of the physicians would prescribe a treatment that is a placebo simply because the patient wanted treatment, 58% would not, and for the remaining 18%, it would depend on the circumstances.

We now know there is a placebo effect. An important task is to obtain the "true" treatment effect from the total effect by subtracting the placebo effect. This can be done in an experiment with two groups in a clinical trial, one group treated with the test drug and the other treated with a placebo. More generally, the placebo group can be replaced with any treatment group (reference group) for comparison. Such a general group is called the control or control group. By subtracting the effect in the placebo group from the test group, we can tell the "pure" effect of the drug candidate. However, such a simple consideration of getting pure treatment effect appears to be naive because knowledge of which treatment a patient is assigned to can lead to subjective and judgmental bias by the patients and investigators. For example, patients who know they are taking a placebo may have the nocebo effect and those who are aware that they are taking the test drug may have overreported the response. To avoid such bias we can implement another experimental technique, called blinding.

Blinding can be imposed on the investigator, the experimental subjects, the sponsor who finances the experiment, or any combination of these actors. In a *single-blind* experiment, the individual subjects do not know whether they have been assigned to the experimental group or the control group. Single-blind experimental design is used where the experimenters must know the full facts (for example, when comparing sham to real surgery). However, there is a risk that subjects are influenced by interaction with the experimenter—known as the experimenter's bias. In *double-blind* experiments, both the investigator and experimental subjects have no knowledge of the group to which they are assigned. A double-blind study is usually better than a single-blind study in terms of bias reduction. In a *triple-blind* experiment, the patient, investigator, and sponsor are all blinded from the treatment group.

Blind experiments are an important tool of the scientific method, in many fields of research. These include medicine, psychology and the social sciences, the natural sciences such as physics and biology, the applied sciences such as market research, and many others.

It is controversial as to how to determine a drug's pure treatment benefit. If a safe drug is approved for marketing, the benefit a patient gets is the effect of the drug without the subtraction of the placebo effect, because companies cannot market a placebo under current health authority regulation in the U.S. Also, the actual medication administration is unblinded. The benefit of a treatment is expected to be more than that from the blind trials.

3.4 Experiment Design

In terms of the structure or layout, experiments can be categorized into several different types: parallel, crossover, cluster, factorial, and titration trials.

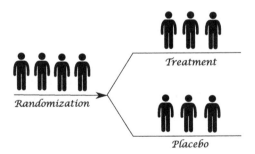

Parallel Design of an Experiment

In a *parallel-group design*, each participant is (usually randomly) assigned to a group to receive interventions (e.g., taking a drug). Subjects assigned to different groups will receive different interventions, e.g., different drugs or the same drug with different doses or dose schedules. A parallel design can be two or more treatment groups with one control group. Parallel designs are commonly used in clinical trials because they are simple, universally accepted, and applicable to acute conditions. A recent study shows that among all clinical trials, 78% were parallel-group trials, 16% were crossover, 2% were cluster, 2% were factorial, and 2% were others.

Unlike in a parallel-group design, in a *crossover design*, each participant is assigned (most often, at random) to a group to receive a sequence of interventions over time. Subjects assigned to different groups will receive different sequences of interventions. There are different crossover designs, the 2×2 crossover design being most commonly used. In a 2×2 design, one group of patients first receives the treatment followed by placebo, and the other group of patients receives the placebo first, followed by the treatment.

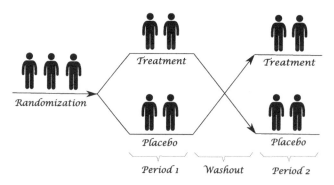

Crossover Design of an Experiment

A crossover study has two advantages: (1) a smaller sample size compared with a parallel design and (2) a reduced influence of confounders because each crossover patient serves as his or her own control. There are four different effects we should note: treatment, sequence, carryover, and period effects.

The *sequence effect* refers to the fact that the order in which treatments are administered may affect the outcome. An example might be that a drug with many adverse effects is given first, making patients taking a second, less harmful medicine more sensitive to any adverse effect.

The *carryover effect* between treatments refers to the situation in which the effect of treatment in the first period carries over to the second period, which confounds the estimates of the treatment effects. In practice, carryover effects can be reduced or eliminated by using a sufficiently long "wash-out" period between treatments. A typical washout period is no less than five times the so-called half time (at which the drug concentration reduces to 1/2 of the maximum concentration in the blood).

The *period effect* refers to the situation in which the same drug given at different time/period will have different effect. A period effect may exist if, e.g., period one is always in the morning or at one clinic and period two is always in the afternoon or at another clinic or home.

In a *cluster randomization trial*, pre-existing groups of participants (e.g., villages, schools) are randomly selected to receive an intervention. A cluster randomized controlled trial is a type of randomized controlled trial in which groups of subjects (as opposed to individual subjects) are randomized. Cluster randomized controlled trials are also known as *group-randomized trials* and *place-randomized trials*.

In a *factorial design* two or more treatments are evaluated simultaneously through the use of varying combinations of the treatments (interventions). The simplest example is the 2×2 factorial design in which subjects are randomly allocated to one of the four possible combinations of two treatments, say, A and B: A alone, B alone, both A and B, neither A nor B. In many cases this design is used for the specific purpose of examining the interaction of A and B. If the number of combinations in a full factorial design is too high to be logistically feasible, a fractional factorial design may be done, in which some of the possible combinations (usually at least half) are omitted.

In a *titration* or *dose-escalation* design, the dose increases over time until reaching the desired level or maximum tolerable level for that patient. There are many other designs we have not discussed here.

3.5 Retrospective and Prospective Studies

Epidemiological studies can be retrospective or prospective. Suppose we want to investigate the relationship between smoking and lung cancer. We can take one of the following approaches: (1) look at hospital records, get the same number of patients with and without lung cancer, then identify them as smokers ("case") or nonsmokers ("control"); (2) look at hospital records, get about the same number of historical smokers and nonsmokers, and find out if they have died of lung cancer or for other reasons, or are still alive; (3) look at hospital records, get about same number of historical smokers and nonsmokers who are still alive, then follow up for a period, say of 10 years, to see if they die of lung cancer or not (or die for other reasons or are still alive). Approaches 1 and 2 are *retrospective studies*; approach 1 is also called a *case-control study*, while approach 2 is called a *historical cohort study*; approach 3 is called a *prospective (cohort) study*.

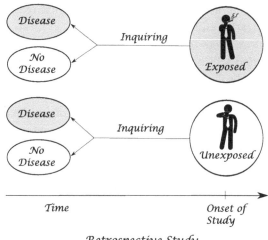

Retrospective Study

As Roger Kirk (1995) states: "Retrospective and prospective studies are observational research strategies in which the independent and dependent variables occur before or after, respectively, the beginning of the study. *Retrospective studies* use historical records to look backward in time; prospective studies look forward in time. A *retrospective study* is particularly useful for studying the relationship between variables that occur infrequently or variables whose occurrence is difficult to predict. A retrospective study also is useful when there is a long time interval between a presumed cause and effect."

Two retrospective studies, one in 1947 by Sir Richard Doll and one in 1951 by A.B. Hill, are considered classics. Doll used the case-control study method and compared the smoking history of a group of hospitalized patients with lung cancer with the smoking history of a similar group without lung cancer. He found 1296 out 1357 Cases (lung cancer patients) were smokers and 61 out of 1357 Control (no lung cancer) were nonsmokers.

Hill used a cohort study, mailing questionnaires to physicians. He categorized them according to their smoking histories and then analyzed the causes of death among those who had died to see whether cigarette smokers had the higher incidence of lung cancer. Over the ten-year period, 4,957 men physicians in the cohort died, 153 of them from lung cancer. The mortality rate per 1000 person-years was 1.30 for smokers and 0.07 for nonsmokers.

Neither the retrospective cohort study nor the case-control study can establish a causal relationship. However, the research strategies can suggest interesting relationships that warrant experimental investigation.

In a retrospective cohort study, more than one dependent variable can be investigated, but only one independent variable can be studied at a time. In a case-control study, multiple independent variables can be investigated, but only one dependent variable can be studied at a time.

In a prospective study (also called a follow-up study, longitudinal study, or cohort study), the data on the independent and dependent variables are collected after the onset of the investigation. Subjects are classified as exposed or nonexposed based on whether they have been exposed to a naturally occurring independent variable. The exposed and unexposed groups are then followed for a period of time, and the incidence of the dependent variable is recorded. The difference between experimental and prospective studies is that in an experimental study, the independent variables are controlled/balanced through a randomization procedure, whereas in a prospective study the independent variables occur naturally.

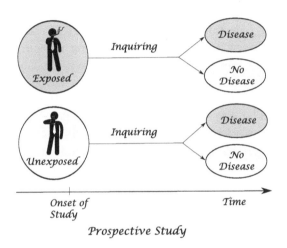

Prospective Study

A well-known example of prospective cohort studies is the study of 22,707 Chinese men set up in Taiwan to investigate the association *between* the primary hepatocellular carcinoma (PHC) among carriers of hepatitis B surface antigen (HBsAg). The study was conducted among male government employees who were enrolled through routine health care services. All participants completed a health questionnaire and provided a blood sample at the time of their entry into the study. Participants were then followed up for an average of 3.3 years. The study results showed that the incidence of PHC among carriers of HBsAg is much higher than among noncarriers (1158/100000 versus 5/100000). The findings support the hypothesis that hepatitis B virus has a primary role in the aetiology of PHC (Beasley et al., 1981).

Prospective studies have advantages over retrospective studies. For instance, the amount of information is not limited by the availability of historical records, and the purported cause (independent variable) clearly precedes the effect (dependent variable). However, prospective studies have some limitations. Mainly, the investigation of a chronic process using a prospective study may require a long follow-up period, which suffers the problems of patients' withdrawals from the study. In the case of a rare event, a sufficient number of subjects may be difficult to obtain.

In addition to prospective and retrospective studies, which look at individuals over a period of time, cross-sectional studies look at a group of people at a given period of time. For example, a study may conduct a questionnaire to determine the prevalence of cigarette smoking.

3.6   Validity and Integrity

In laboratory testing, *validity* is the accuracy of a test, which is dependent on the instrument, the person who conducts the test, and the lab environment. *Reliability* is how repeatable the test result is. A test can be valid, but the reliability can be very low. For example, one can measure blood pressures very accurately, but each of such measures can be not reliable if the patients' blood pressures vary a lot each time. In this case, maybe the average of the blood pressures over time is valid and more reliable. Here we are considering not only the valid lab test, but also more broad studies, such as clinical trials.

Two goals of research are to draw valid conclusions about the effects of an independent variable and to make valid generalizations to populations. Adapted from Kirk (1995), three different categories of threats to these goals are considered:

(1) *Statistical conclusion validity* is concerned with threats to valid inference making that result from random error and the ill-advised selection of statistical procedures.
(2) *Internal validity* is concerned with correctly concluding that an independent variable is, in fact, responsible for variation in the dependent variable.
(3) *External validity* is concerned with the generalizability of research findings to and across populations of subjects and settings.

Threats to internal validity include the following difficulties:

(1) Repeated testing of subjects may result in familiarity with the testing situation or acquisition of information that can affect the dependent variable.
(2) Changes or calibration of a measuring instrument can cause a change in accuracy and precision of the instrument and affect the measurement of the variable.
(3) Placebo and nocebo effects can occur as discussed earlier.
(4) Screening, especially multiple screenings to qualify subjects to meet the inclusion criteria, may distort the response because the baseline value may not really reflect the baseline. On the other hand, if the baseline value has

a large variability, multiple screenings may be necessary to get an average baseline value.

(5) Sample selection criteria are changing over time. For example, in clinical trials protocol amendments are not unusual when the recruitment is slow. Such a change in entry criteria may cause extra confounding variables that are not easy to recognize.

(6) Early study termination from the experiment or other reasons that cause missing or incomplete observations and missing data will cause difficulties in statistical analysis and could lead to a considerable bias in some cases.

Unbiasedness *Biasedness*

Integrity and Validity

With regard to the threats to external validity, if (1) internal validity holds, (2) the experimental sample is representative of the intended population, and (3) experimental conditions are the same (similar) as outside reality, then we can extrapolate the experimental result to the population. However, checking (2) and (3) is often difficult. This is because, for instance, the experiment may be conducted early and results may be used at a later time. Such cases often occur in clinical trials when the to-be-approved drugs will be used for many years. During the time a drug first comes on market and is available to the doctors' prescription, many things can change (e.g., the standard care may be improved). The results obtained under conditions of repeated testing may not generalize to situations that do not involve repeated testing and vice versa. Even samples obtained from multiple screenings may not be representative of the target population.

A controversial issue surrounding internal and external validity is so-called subgroup analysis. Taking the clinical trial as an example, with economic globalization there are ever more international clinical trials (multiregional trials). Suppose the drug has demonstrated efficacy in some countries but not in others, while the drug is seen to be effective when combining all countries into one entity. Then to what extent should we claim the drug is effective, in some of the countries, in all the countries in the trial, or in all the countries in the world?

3.7   Confounding Factors

In statistics, a *confounding factor* (also *confounder, hidden variable,* or *lurking variable*) is an extraneous variable that correlates, positively or negatively, with both the dependent variable and the independent variable. Such a relationship between two observed variables is termed a spurious relationship.

A classic example of confounding is to interpret the finding that people who carry matches are more likely to develop lung cancer as evidence of an association between carrying matches and lung cancer. Carrying a match or not is a confounding factor in this relationship: smokers are more likely to carry matches and they are also more likely to develop lung cancer. However, if "carrying matches" is replaced with "drinking coffee," we may easily conclude that coffee more likely causes cancer. Interestingly enough, epidemiological studies have evaluated the potential association between coffee consumption and lung cancer risk. The results were inconsistent. Recently Tang et al. (2010) conducted a meta-analysis of eight case-control studies and "confirmed" that coffee consumption is a risk factor for lung cancer.

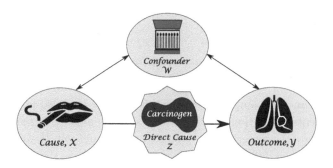

Direct Cause versus Confounder

For a variable to be a confounder in a clinical trial it must satisfy three conditions:

(1) It must be associated with the treatment (the main factor).
(2) It must be a predictor (not necessarily a cause) of the outcome being measured.
(3) It must not be a consequence of the treatment (the main factor) itself.

A factor is not a confounder if it lies on the causal pathway between the variables of interest. Such a factor can be a surrogate or direct cause. For example, the relationship between diet and coronary heart disease may be explained by measuring serum cholesterol levels. Cholesterol is a surrogate, but not a confounder because it may be the causal link between diet and coronary heart disease. In cigarette-smoke-caused lung cancer, carcinogen in the cigarette is the direct cause of the cancer.

Bias creates an association that is not true, while confounding describes an association that is true, but potentially misleading. Confounders are more usually a problem in observational studies, where the exposure of a risk factor is not randomly distributed between groups. In evaluating treatment effects from observational data, prognostic factors may influence treatment decisions, producing the so-called confounding-by-indication bias. Controlling for known prognostic factors using randomization can reduce such a bias, but it is always possible that unknown confounding factors were not balanced even when a randomization procedure was used.

In terms of their source, there are operational confounds, which are due to the choice of measurement instrument; procedural confounds due to situational characteristics; and person confounds, which occur because of interindividual differences.

An *operational confound* is a type of confound that can occur in both experimental and observational study designs when a measure designed to assess a particular construct inadvertently measures something else as well.

A *procedural confound* is a type of confound that can occur in an experiment when the researcher mistakenly allows another variable to change along with the manipulated independent variable.

In a clinical trial, some adverse reactions (ARs) may occur randomly and depend on the length of the follow-up period. The longer the follow-up is, the more ARs one is likely to see. When two drugs in evaluation have different dose schedules (e.g., one is given weekly, the other is given biweekly), this may lead to a different observational period and/or frequency for ARs, and artificially make one drug appear to be safer than the other, even if in fact they have a similar safety profile. Likewise, consider the case when drug effectiveness is measured by the maximum response during the treatment period. If two drugs in evaluation, one given once a week, the other given 5 times a week, have their effects on patients measured in the clinic, the difference in observational frequency can create an artificial treatment difference simply because drug effects include a random component, so that more frequent measurements of response are more likely to capture a maximum value.

A confounding factor, if measured, often can be identified and its effect can be isolated from other effects using appropriate statistical procedures. A recent example is an ABC poll (http://abcnews.go.com, 2004) that found that Democrats were less satisfied with their sex lives than Republicans. But women are also less satisfied with their sex lives than men, and more women are Democrats than Republicans. How do we know whether the correlation between happy sex lives and political affiliation is an honest relationship, or just a side effect of the gender differences between Democrats and

Republicans? A statistical "adjustment" for the confounding factor (in this case, gender) would solve the problem.

To control confounding in an experiment design, we can use randomization or stratified randomization. The latter will provide a better balance of confounders between intervention groups, and confounding can be further reduced by using appropriate statistical analyses such as the so-called analysis of covariance.

3.8   Variation and Bias

Variations can come from different sources. *Variability within an individual* is the variation in the measures of a subject's characteristics over time. *Variability between individuals* concerns the variation from subject to subject. Instrumental variability is related to the precision of the measurement tool. Other variabilities can be attributed to the difference in experimenter or other factors.

Variability in measurements can be either random or systematic. *Bias* is a systematic error in a study that leads to a distortion of the results. Human perception occurs by a complex, unconscious process of abstraction, in which certain details of the incoming sense data are noticed and remembered, and the rest are forgotten. What we keep and what we throw away depends on an internal model or representation of the world that is built up over our entire lives. The data is fitted into this model. Later when we remember events, gaps in our memory may even be filled by "plausible" data our mind makes up to fit the model (wikipedia.org). The concepts of variation and bias are illustrated in the figure below.

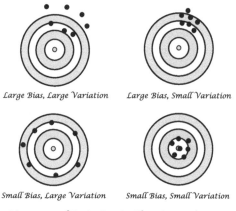

Large Bias, Large Variation   Large Bias, Small Variation

Small Bias, Large Variation   Small Bias, Small Variation

Bias versus Variation in Shooting a Target

There are other sources of bias, such as selection bias, ascertainment or confirmation bias, and publication bias.

Selection bias refers to a subject selection process or sampling procedure that likely produces a sample not representative of the population, or produces samples that result in systematical (probabilistical) imbalance of confounding variables between comparison groups. For instance, a student is conducting an experiment to test a chemical compound's effect on mice. He knows that inclusion of a (placebo) control and randomization are important, so he randomly catches a half of the mice from the cage and administers the test compound. After days of observations, he finds that the test compound has a negative effect on mice. However, he might overlook an important fact: the mice he caught were mostly physically weaker and thus they were more easily caught. Such selection bias can also be seen in the capture–recapture method due to physical differences in animals.

Another example would be the following: A clinical trial investigator might use her knowledge to assign treatments to different patients, say, sicker patients go to the new treatment group and less sick patients to the control group. Such a treatment assignment procedure may exaggerate or understate the treatment difference. Selection bias can occur in surveys too. For instance, in an Internet survey, selection bias could be caused by Internet accessibility or by the differences in familiarity with a computer. Randomization can reduce bias but cannot completely prevent it, especially when there is a larger number of potential confounding factors.

Confirmation bias is a tendency of people to favor information that confirms their beliefs or hypotheses. Confirmation bias is best illustrated using examples in clinical trials. When knowledge of the treatment assignment can affect the objective evaluation of treatment effect and lead to systematic distortion of the trial conclusions, we have what is referred to as observer or ascertainment bias.

A patient's knowledge that he is receiving a new treatment may substantially affect the way he feels and his subjective assessment. If a physician is aware of which treatment the patient is receiving, this can affect the way he collects the information during the trial and influence the way the assessor analyzes the study results.

Confirmation bias happens more often and is more severe in subjective evaluations than in "objective laboratory evaluations." However, it does exist in laboratory-related outcomes, where the knowledge of treatment assignment can have an impact on how the test is run or interpreted, although the impact of this is most severe with subjectively graded results, such as pathology slides and radiological images. Blinding can effectively reduce ascertainment bias as discussed in Section 3.3.

Human observations are biased toward confirming the observer's conscious and unconscious expectations and view of the world; we "see what we expect or want to see." This is called confirmation bias. This is not deliberate falsification of results, but it can happen to good-faith researchers. As a result, people gather evidence and recall information from memory selectively, and interpret it biasedly.

How much attention the various perceived data are given depends on an internal value system, which judges how important they are to us. Thus, two people can view the same event and come away with entirely different perceptions of it, even disagreeing about simple facts. This is why eyewitness testimony is so unreliable.

Confirmation bias is remarkably common. Whenever science meets some ideological barrier, scientists are accused of, at best, self-deception or, at worst, deliberate fraud. We should know that one can tell all the truth, and nothing but the truth, but at the same time what one says can still be very biased. For example, you might talk about only the things on the positive side of the ledger. In a church or other religious society, the same stories are told again and again; these stories often fit only into the society's needs.

Similarly, *publication bias* is the tendency of researchers and journals to more likely publish results that appear positive, but they may or may not be truly positive. Some publication bias can be avoided or reduced, but other instances of it cannot. On the other hand, we cannot publish positive and negative results equally; otherwise, since there are a lot more negative results than positive ones, it may be too much for our brains to handle!

Bias can be caused by incompetent questionnaires. A teacher believes that attendance is an important aspect in learning; hence, he keeps a record of his students who miss class. Every day he asks before class starts: "For those who are absent please raise your hand." Of course, no one raises his hand because those who are absent cannot hear it. For that reason he is ecstatic about his students' "excellent attendance." One might think that this is a silly example, but similar things have happened in reality: The HR of a company conducted a company-wide employee survey as a basis for management improvement. One of the multiple choice questions was: "Did your manager consider your suggestions seriously?" The options were: (1) very seriously, (2) somewhat seriously, (3) not at all, and (4) I don't know. The survey result was very positive: most employees chose the answer "very seriously." However, later HR accidentally discovered that the employees didn't want to ask why the company didn't keep offering free Coca Cola even though they really liked the free drink. HR's further investigation indicated that this was a broader issue than just the Coca Cola instance; most employees only made a suggestion when they were pretty sure their managers would like it and would take action on it.

Missing observations or *incomplete data* can also cause bias in data analysis, especially when the missing mechanism is not random. Consider, for example, that in clinical trials, missing data often results from patients' early termination, due to ineffectiveness of the treatment or to safety issues. Missing data can distort available data and make statistical analysis difficult.

When encountering missing data, or patients' noncompliance to the study protocol (e.g., a patient who takes too little or too much of a drug), the experimenter may choose to exclude some data. Such subjective exclusion of data can also introduce bias.

Model selection can introduce bias too. People tend to select a simpler model even when a complicated model might explain things a little better. The selection of a simpler model can be made in the name of parsimony or mathematical simplicity. This could lead to a special type of bias—parsimony or *model selection bias.*

There are other sources of bias. For instance, when multiple screenings are used to qualify subjects for an experiment, bias can be introduced because the qualifying screening value may just happen to be lower than it actually should be and after subjects are entered into the study, the value can come back to the natural status without any intervention. Such a phenomenon is called "regression to the mean." Such bias can be removed by adding a control group and randomization.

In statistics, the bias of an estimator for a parameter is the difference between the estimator's expected value and the true value of the parameter being estimated. It is a common practice for a frequentist statistician to look for the minimum unbiased estimator. However, it is not always feasible to do so, and unbiased estimators may not even exist. Therefore, we have to balance between variability and bias.

Finally, let me quote myself: "Each of us is biased in one way or another. However, collectively, we as a social entity are not biased, which must be. In fact, the collective view defines 'reality' or the unbiased world."

3.9   Randomization

Randomization is a procedure to assign subjects or experimental units to a certain intervention based on an allocation probability rather than by choice. For example, in a clinical trial, patients can be assigned one of two treatments available with an equal chance (probability 0.5) when they are enrolled in the trial.

The utilization of randomization in an experiment minimizes selection bias, balancing both known and unknown confounding factors, in the assignment of treatments (e.g., placebo and drug candidate). In a clinical trial, appropriate use of randomization procedure not only ensures an unbiased and fair assessment regarding the efficacy and safety of the test drug, but also improves the quality of the experiments and increase the efficiency of the trial.

Randomization procedures that are commonly employed in clinical trials can be classified into two categories: conventional randomization and adaptive randomization.

Conventional randomization refers to any randomization procedures with a constant treatment allocation probability. Commonly used conventional randomization procedures include simple (or complete) randomization, stratified randomization, cluster randomization, and block randomization.

A *simple randomization* is a procedure that assigns a subject to an experimental group with a fixed probability. Such a randomization probability can be different for different groups, but the sum of the randomization probabilities should be 1. For example, in a clinical trial, patients can be assigned to one of two experiment groups with probability 0.5. Such a trial is said to have balanced design because the sample sizes (number of subjects) in the two groups are expected to be the same.

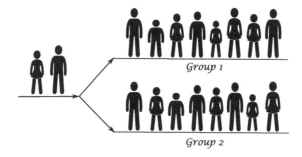

Simple Randomization

Like simple randomization, a *stratified randomization* is a randomization with a fixed allocation probability. When there are important known covariates (confounding factors), stratified randomization is usually recommended to reduce treatment imbalances. For a stratified randomization, the target population is divided into several homogenous strata, which are usually determined by some combinations of covariates (e.g., patient demographics or patient characteristics). In each stratum, a simple randomization is then employed.

Cluster randomization is another conventional randomization. In certain trials, the appropriate unit of randomization may be some aggregate of individuals. This form of randomization is known as cluster randomization or group randomization. Cluster randomization is employed by necessity in trials in which the intervention is by nature designed to be applied at the cluster level such as with community-based interventions.

In practice, conventional randomization procedures could result in severe imbalance in the sample size or in confounding factors between treatment groups at some point during the trial. When there is a time-dependent heterogeneous covariance that relates to treatment responses, it could induce a bias in the results. A *block randomization* is used to reduce the imbalance over time in treatment assignments, which may occur when a simple randomization is used. In block randomization, the allocation probability is a fixed constant before either of the two treatment groups reaches its target number. However, after the target number is reached in one of the two treatment groups, all future patients in the trial will be assigned to the other treatment group.

Adaptive randomization is a randomization technique in which the allocation of patients to treatment groups is based on the observed treatment effect: The treatment group showing a larger effect will have a larger probability of assigning newly enrolled subjects. In this way, the probability of a new subject being assigned the better treatment will gradually increase. A commonly used response-adaptive randomization is the so-called randomly-play-the-winner (Chang 2007).

Theoretically, adaptive randomization can cause the so-called accrual bias, whereby volunteers may try to be recruited later in the study to take advantage of the benefit from previous outcomes. Earlier enrolled subjects have higher probabilities of receiving the inferior treatment than later enrolled subjects.

3.10 Adaptive Experiment

A classical experiment design has static features, such as a fixed sample size, fixed randomization, and fixed hypothesis tests. In contrast, an adaptive experimental design allows for changing or modifying the characteristics of an experiment based on cumulative information. These modifications are often called adaptations. The word *adaptation* is so familiar to us because we make adaptations constantly in our daily lives according what we learn over time. Some of the adaptations are necessary for survival, while others are made to improve our quality of life. We can be equally smart in conducting our research by making adaptations based on what we learn as a study progresses. These adaptations are made because they can improve the efficiency of the experimental design, provide earlier remedies, and reduce the time and cost of research. An adaptive design is also ethically important in clinical trials. It allows for stopping a trial earlier if the risk to subjects outweighs the benefit, or when there is early evidence of efficacy for a safe drug. An adaptive design may allow for randomizing more patients to the superior treatment arms and reducing exposure to inefficacious, but potentially toxic, doses. An adaptive design can also be used to identify better target populations through early biomarker responses.

Adaptive designs have become very popular in drug development in recent years because of the increasing cost and failure rate in drug development. Reasons for this challenge include the following: (1) a diminished margin for improvement escalates the level of difficulty in proving drug benefits, (2) genomics and other new sciences have not yet reached their full potential, and (3) easy targets are the focus as chronic diseases are more difficult to study (Woodcock, 2005). An improvement in methodology is urgently needed, and adaptive designs fit the need perfectly.

An *adaptive design* is a clinical trial design that allows adaptations or modifications to aspects of the trial after its initiation without undermining the validity and integrity of the trial (Chow and Chang, 2006). There are specific statistical methodologies that are required for an adaptive design to retain validity and integrity (Chang 2007). While there are different types of adaptive designs, they feature a common characteristic, i.e., interim analysis, which is a statistical analysis performed on partial (interim) data. The results of the interim analysis will serve as the basis for the adaptations in the ongoing trial.

A *group sequential design* is an adaptive design that allows for premature termination of a trial due to efficacy or futility, based on the results of interim analyses on early collected partial data. For a trial with a positive result, early stopping ensures that a new drug product can be exploited sooner. If a negative result is indicated, early stopping avoids wasting resources. Sequential methods typically lead to savings in sample size, time, and cost when compared with the classic design with a fixed sample size

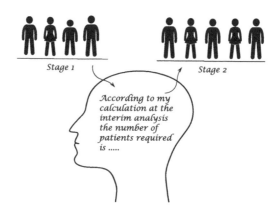

Adaptive Experiment with Sample Size Re-Estimation

A *sample size re-estimation design* refers to an adaptive design that allows for sample size adjustment or re-estimation based on the review of interim analysis results. The sample size requirement for a trial is sensitive to the treatment effect and its variability. An inaccurate estimation of the effect size and its variability could lead to an underpowered or overpowered design, neither of which is desirable. Sample size re-estimation allows us to reset the sample size to retain the targeted power.

A *drop-loser design* is an adaptive design consisting of multiple stages. At each stage, interim analyses are performed and the losers (i.e., inferior treatment groups) are dropped from the experiment. Ultimately, the best arm(s) are retained. If there is a control group, it is usually retained for the purpose of comparison.

An *adaptive randomization design* is a design that allows modification of allocation probabilities (i.e., the probabilities of assigning patients to different treatment groups) during the conduct of the trial as we discussed in Section 3.9. The allocation probability is based on the response of the previous patients. When more responders are observed from one treatment group, the probability of assigning the next patient to that group increases so that gradually more patients will be assigned to the superior group. Adaptive randomization was initially proposed because of ethical considerations, i.e., it tends to increase the likelihood of benefit to patients in a study.

A *biomarker-adaptive design* is a design that allows for adaptations using information obtained from biomarkers (e.g., tumor response), for which the response can be measured earlier and/or more easily than the ultimate clinical endpoint such as survival. A biomarker is a characteristic that is objectively measured and evaluated as an indicator of normal biologic or pathogenic processes or pharmacological response to a therapeutic intervention (Chakravarty, 2005). A biomarker can be a classifier, prognostic, or predictive marker.

A *classifier biomarker* is a marker that usually does not change over the course of the study, for example, DNA markers. Classifier biomarkers can be used to select the most appropriate target population, or even for personalized treatment. Classifier markers can also be used in other situations. For example, it is often the case that a pharmaceutical company has to make a decision whether to target a selective population for whom the test drug likely works well or to target a broader population for whom the test drug is less likely to work well. However, the size of the selective population may be too small to justify the overall benefit to the patient population. In this case, a biomarker-adaptive design may be used, where the biomarker response at interim analysis can be used to determine which target populations should be the focus of our attention.

3.11   Ethical Issues

Ethical issues are so very important to everyone who is involved in research. As Kirk (1995) pointed out: "In recent years, the research community has witnessed a renewed resolve to protect the rights and interests of humans and animals. Codes of ethics for research with human subjects have been adopted by a number of professional societies. In addition to codes of ethics of professional societies, legal statutes, and peer review, perhaps the most important regulatory force within society is the individual researcher's ethical code. Researchers should be familiar with the codes of ethics and statutes relevant to their research areas and incorporate them into their personal codes of ethics."

A way of defining *ethics* focuses on the disciplines that study standards of conduct, such as philosophy, theology, law, psychology, and sociology. For example, a "medical ethicist" is someone who studies ethical standards in medicine.

According to Dr. Resnik at the National Institutes of Health (NIH.gov), there are several reasons why it is important to adhere to ethical norms in research. First, norms promote the aims of research, such as knowledge, truth, and avoidance of error. Second, since research often involves a great deal of cooperation and coordination among many different people in different disciplines and institutions, ethical standards promote the values that are essential to collaborative work, such as trust, accountability, mutual respect, and fairness. Third, many of the ethical norms help to ensure that researchers can be held accountable to the public. Fourth, ethical norms in research also help to build public support for research; people are more likely to fund a research project if they can trust the quality and integrity of research.

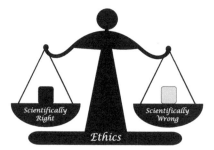

Ethical versus Scientific Standards

The following Codes and Policies for Research Ethics are adapted from Shamoo and Resnik (2009) at the National Institute for Health (NIH.gov, 2009):

Honesty: Strive for honesty in all scientific communications. Honestly report data, results, methods and procedures, and publication status. Do not fabricate, falsify, or misrepresent data. Do not deceive colleagues, granting agencies, or the public.

Objectivity: Strive to avoid bias in experimental design, data analysis, data interpretation, peer review, personnel decisions, grant writing, expert testimony, and other aspects of research where objectivity is expected or required. Avoid or minimize bias or self-deception. Disclose personal or financial interests that may affect research.

Integrity: Keep your promises and agreements; act with sincerity; strive for consistency of thought and action.

Carefulness: Avoid careless errors and negligence; carefully and critically examine your own work and the work of your peers. Keep good records of research activities, such as data collection, research design, and correspondence with agencies or journals.

Openness: Share data, results, ideas, tools, resources. Be open to criticism and new ideas.

Respect for Intellectual Property: Honor patents, copyrights, and other forms of intellectual property. Do not use unpublished data, methods, or results without permission. Give credit where credit is due. Give proper acknowledgement or credit for all contributions to research. Never plagiarize.

Confidentiality: Protect confidential communications, such as papers or grants submitted for publication, personnel records, trade or military secrets, and patient records.

Human Subjects Protection: When conducting research on human subjects, minimize harms and risks and maximize benefits; respect human dignity, privacy, and autonomy; take special precautions with vulnerable populations; and strive to distribute the benefits and burdens of research fairly.

Other ethical priorities include the practice of professional responsibilities, such as responsible publication, mentoring, and respect for colleagues, as well as social responsibility, nondiscrimination, legality, and humane care for animal subjects.

The Animal Welfare Act promotes animal rights. The so-called 3Rs are well-known guiding principles in animals testing (Russell and Burch, 1959): (1) **R**eplacement of animal methods by of nonanimal methods whenever it is possible to achieve the same scientific aims, (2) **R**eduction in animal use to obtain comparable levels of information, and (3) **R**efinement of test methods to minimize potential pain, suffering, or distress and enhance animal welfare for the animals used.

The boiling frog story is a widespread anecdote: if a frog is slowly being boiled alive, it will not perceive the danger and will not try to escape but be cooked to death. However, there are several experiments with conflicting results, some confirmatory while others negative. Now here is a question for you: How do you find out if "boiling frog syndrome" is true or not without cooking a frog to death?

Chapter 4

Scientific Inference

4.1 The Concept of Probability

Probability is one of the most important concepts in modern science. It is interesting and frustrating to know that there are two different notions of probability, frequentist and Bayesian. However, I will illustrate (Section 6.11) that the difference between these two interpretations is found only on the surface or in a calculational aspect, and is not fundamental. If the two conceptions of probability are fundamentally different, how can Bayesian and frequentist statisticians and the general public communicate with each other? Why do they continue to use the same words for different meanings?

The *probability* of an event is the ratio of the number of cases of interest to the number of elementary events in the sample space. A sample space is the set of all possible outcomes, whose elementary events have the same probability (or possibility). The equal possibility is the consequence of the principle of indifference, which states that if the n mutually exclusive, independent possibilities are indistinguishable except for their names, then each possibility should be assigned a probability equal to $1/n$. You see, we use possibility to define elementary events and use elementary events to define probability—a circular definition! But I don't think we can do much better than that. As we have discussed in Chapter 1, the essence of understanding is concept-mapping which is eventually a circular definition however you look at it.

What is the probability of heads when one throws a fair coin? Your answer might be 0.5 due to the indifference principle. In this case the sample space (all possible outcomes) is {head, tail} and the two elementary events {head} and {tail} have the equal probability 0.5. If we concern ourselves with the outcome when one throws a die, the sample space will be $\{1, 2, 3, 4, 5, 6\}$, and its elementary events all have the same probability $1/6$. However, if we toss two dies, what is the probability of getting the sum of the two outcomes equal to 3? There are $6 \times 6 = 36$ possible outcomes in the samplespace. Two of them are of interest: $\{1, 2\}$ and $\{2, 1\}$. Therefore the probability of having the sum equal to 3 is 2/36.

Statisticians often say that a random variable (a variable whose value is random) X comes from a Bernoulli distribution with parameter or rate of success 0.3. This means the probability of observing $x = 1$ is 0.3. Here the term *probability* means that when the "same" experiment is repeated many times the proportion of successes will approach 30%. When we say that x is drawn or generated from probability distribution $f(x)$, we imply that if the experiment is repeated in the long run, the relative frequencies of random numbers will approximate the distribution $f(x)$.

There are two problems with the probability definition that is based on the basis of repeating the same experiment a number (infinite) of times:

(1) The same experiments cannot be exactly the same. If they are exactly the same, then the outcome of each experiment will *be* the same. For example, in the coin tossing or dice experiment, we didn't specify from which height we would drop the object. In fact, each time it will be different. Therefore, the "same experiment" actually means similar experiment. However, we have not defined what we mean by the term *similar*.

(2) The fuzziness of "sameness" in the above definition of probability is both good and bad for the calculation of probabilities. The good thing is that we can always find a set of repeated experiments by "tuning" the meaning of the word *similar* for the purpose of calculating the probability. The bad thing is that the probabilities as calculated by different people can be different because of the difference in the concept of *similarity* results in different sets of experiments.

I toss a Chinese coin once and ask: "What is the probability the coin comes up heads?" Either heads or tails, the result is fixed but unknown. In this case, we can define the probability as follows: flip the same Chinese coin under similar conditions many (infinite) times, the proportion of heads is the probability.

A *frequentist* statistician may proceed like this: "Since I don't have the experimental data, I have to flip the Chinese coin and calculate the proportion of heads."

The *Bayesian* calculation goes something like this: "Since U.S. Coins are similar to Chinese coins, I use the probability for flipping U.S. coins for the estimation until the data of flipping Chinese coins starts accumulating. More precisely, the combination of my prior knowledge about U.S. coins and the data from the experiment on Chinese coins produce, the probability of interest."

Both frequentist and Bayesian statisticians will come to the same conclusion when the experiment is conducted "infinitely many" times. Therefore, what the statisticians from the two different statistical paradigms ultimately look for (i.e., the definition of probability) is the same, i.e., the proportion of favorable outcomes among all possible outcomes, but the calculations of the probability are different over time.

The notion of Bayesian probability is a measure of the state of individual knowledge about a hypothesis. This view is certainly not fully operational or specific enough for calculating the probability (subjective of prior), but it is consistent with what we have in dealing with things in our daily lives.

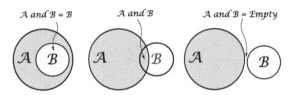

Joint Event of A and B

The *joint probability* of two events A and B, denoted by $\Pr(A \text{ and } B)$ is the probability that the two events both occur. For instance, in the experiment of flipping two coins, if A represents "the first coin is showing heads" and B represents "the second coin is showing heads," then the joint probability of both heads is $\Pr(A \text{ and } B) = 0.5 \times 0.5 = 0.25$. This is because A and B are two *independent events*, which means statistically $\Pr(A \text{ and } B) = \Pr(A)\Pr(B)$.

Naturally, we may ask: What is the probability that either A or B occurs? This probability is denoted by $\Pr(A \text{ or } B)$. The probability is calculated using the formulation:

$$\Pr(A \text{ or } B) = \Pr(A) + \Pr(B) - \Pr(A \text{ and } B).$$

For instance, the probability of either of the two coins showing heads is $\Pr(A \text{ or } B) = 0.5 + 0.5 - 0.5 \times 0.5 = 0.75$.

Another basic probability concept regarding multiple events is the so-called *conditional probability*. Given two events A and B with $\Pr(B) > 0$, the conditional probability of A given B, written as $\Pr(A|B)$, is defined as the quotient of the joint probability of A and B, and the probability of B:

$$\Pr(A|B) = \frac{\Pr(A \text{ and } B)}{\Pr(B)}.$$

It is usually easier to obtain the conditional probability than the joint probability, therefore, the joint probability can be calculated using $\Pr(A \text{ and } B) = \Pr(A|B)\Pr(B)$.

In practice, we are more interested in a special type of random events, the events whose outcomes are real values or associate with real value. Two trivial examples will be the experiments of die-throwing and coin-flipping with value 1 assigned to heads and 0 to tails. Such events are called random variables. In statistics, dependence refers to any statistical relationship between two random variables. Correlation refers to a broad class of statistical relationships involving dependence. The strength of a correlation is measured by a correlation coefficient, which may refer to Pearson's correlation, Spearman's rank correlation, Kendall's correlation τ, and others.

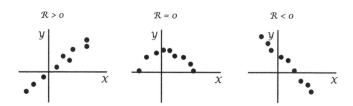

Three Dependent Scenarios

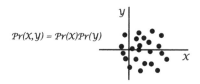

An Independent Scenario ($R = 0$)

Pearson's Correlation R versus Independence

Pearson's correlation coefficient R, a measure of the strength and direction of the linear relationship between two variables, is defined as the (sample) covariance of the variables divided by the product of their (sample) standard deviations. *Spearman's correlation* is defined as the Pearson correlation coefficient between the ranked variables and is a measure of how well the relationship between two variables can be described by a monotonic function. *Kendall's correlation* is a measure of the portion of ranks that match between two data sets.

The absolute value of Pearson correlation coefficients is no larger than 1. Correlations equal to 1 or –1 correspond to data points lying exactly on a straight line. The Pearson correlation coefficient is symmetric, i.e., the correlation between X and Y is the same as that between Y and X.

If two random variables are independent, the correlation coefficient must be zero. However, a zero correlation coefficient does not imply independence. Random variables X and Y are independent if the occurrence of one does not depend on the other. In other words, the joint probability of X and Y is the product of the two probabilities of X and Y, i.e., $\Pr(X \text{ and } Y) = \Pr(X)\Pr(Y)$. Also, bear in mind that correlation does not imply causation as discussed in Chapter 1.

4.2 Probability Distribution

To understand probability distributions, we need to first distinguish between discrete and continuous random variables. In the discrete case, we assign a probability to each possible value: for example, when throwing a die, 1/6 is assigned as the probability to each of the six values 1 to 6. Such a probability function is called a *probability mass function*. In contrast, when a random variable, X, takes real values from a continuum, the so-called *probability density function* $f(x)$ is used, which is, loosely speaking, the probability per unit X. The cumulative distribution function $F(x)$ gives the probability of observing the random variable X having a value no larger than a given value x.

An experiment such as flipping a biased coin can be characterized using the *Bernoulli distribution*, which takes value 1 (heads) with probability p and value 0 (tails) with probability $q = 1 - p$. If one conducts n independent Bernoulli experiments, the number of successes (1's) is characterized by the *binomial distribution*. The Bernoulli and binomial distributions are two common discrete distributions, while the *normal distribution* is an example of a continuous distribution. The normal, denoted by $N\left(\mu, \sigma^2\right)$ is one of the most widely used distributions in probability theory and practice. Here the mean μ indicates the central location of the distribution and the standard deviation σ represents the dispersion of the random variable (see the diagram).

The normal distribution is very common because of the *central limit theorem*, which says that the mean of a sufficiently large number of independent random variables with a finite mean and variance will be approximately normally distributed. Thus, the binomial distribution becomes approximately a normal distribution with mean p and the standard deviation $\sigma = \sqrt{p(1-p)/n}$ when the sample size n is large.

Another common continuous distribution is the *exponential distribution*, which is used for variables that take only positive values such as survival or time to event data. Survival data often involve truncated data or *censoring*. Ideally, both the birth and death dates of a subject are known, in which case the lifetime is known. However, if it is known only that the date of death comes after some date, the resulting data is an example of what is called right censoring. *Right censoring* will occur for those subjects whose birth date is known but who are still alive when they are lost to follow-up or when the experiment ends.

One may have questions about the word *probability* we have used here in the probability distribution. When we talk about the probability

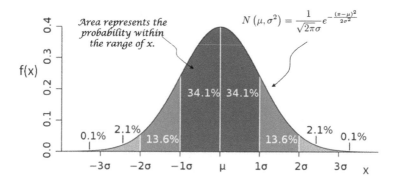

Normal Distribution

distribution $f(x)$, the term *probability* means that if the random sample x
is drawn repeatedly (repeated experiments) from the same source independently, the empirical distribution (relative frequencies) of x will approach to
$f(x)$ when the sample increases. In fact, this is the notion of *computer simulation* or *Monte Carlo* techniques.

There are frequently used terms we should know:

(1) *Expected value*: the weighted average of the possible values, using their
    probabilities as their weights;
(2) *Median*: the value such that the set of values less than the median has a
    probability of one-half;
(3) *Variance*: the standard deviation squared, an important measure of the
    dispersion of the distribution;
(4) *Skewness*: a measure of the symmetry of a distribution; and
(5) *Correlation*: a measure of dependence of between two random variables.

Two common measures for precision of a random variable are the *confidence interval* (CI) and the Bayesian *credible interval* (BCI). For example, the
meaning of a 95% CI in this situation is that if samples are drawn repeatedly
from a population and a CI is calculated for each, then 95% of those CI's
should contain the population parameter. However, we cannot say that the
probability of the true mean falling within a particular sample CI is 95%. In
Bayesian statistics, a credible interval is a probability interval. For example,
a statement such as "a 95% credible interval for the response rate θ is 0.22–
0.60" means that the probability that θ lies in the interval from 0.22–0.60 is
95%.

Knowing the distribution of a random variable of interest is critical for
making scientific inference (about model parameters). Unfortunately in reality, we often donot know the true distribution f of a random variable. What
we often can have is a sample from the population. Fortunately we can treat

the sample as a representative of the population and construct the so-called empirical distribution. In other words, the relative frequency of a sample of n subjects is an approximation of the true distribution f. To make use of the empirical distribution $\hat{f}$, statistical scientists have developed several simulation methods such as bootstrapping, resampling, and permutation methods. These methods have a common notion—*plug-in principle*: The estimate of a parameter $\theta = t(f)$ can be approximated by the plug-in estimate $\theta = t(\hat{f})$, where t represents a function.

4.3 Evidential Measures

The choice of an evidential measure (endpoint, dependent variable) depends on the underlying question. The evidential measure chosen should measure what is supposed to be measured and should be sensitive enough to changes of the factors under study.

Suppose we want to assess the impact of deaths and diseases. If the impact concerns society, the simplest measure is the number of deaths or the number of deaths per year. This is also an overall measure of impact on all people (families) in the society or country. However, this measure may not be appropriate if we want to make comparisons among different societies or age groups since the population sizes are different for different societies and age groups; i.e., a larger number of deaths is expected in a larger population. For this reason, we can use a death rate (e.g., number of deaths per 1000 people per year). Ethically, lives are equally important. Hence, a death, whether of someone young or someone old, has the same impact. On the other hand, a death in young age will mean a longer life expectancy loss than a death in old age. Therefore, one measure is the expected life year loss. But this still does not satisfy our expectation because we not only want to live long, but also want to live better. There is a so-called quality of life (QOL) adjusted life years or QOL-adjusted (weighted) life expectancy. To measure a disease burden we may compare QOL scores in the patient population against the standard population. If we want to measure the impact on the work productivity, we may use work days lost due to disease.

Statistically, evidential measures can be continuous such as blood pressure, binary such as death (yes or no), ordinal as with disease status (worsening, stable, or improving), or others types.

For continuous variables, we often use the mean, median, mean change, percent change from baseline, mean difference between groups, and confidence interval and credible intervals. For a survival or *time-to-event* variable (e.g., lifetime of a light bulb), we use mean or median survival time, or hazard rate, or hazard ratio. For binary variables, we often use proportion, proportion change, proportional difference, ratio, odds, odds ratio, relative risk, and others.

Incidence and Prevalence: Incidence is a measure of the risk of developing some new condition within a specified period of time. The incidence rate is the number of new cases per population in a given time period. If the incidence rate of a disease increases, then there is a risk factor that promotes the incidence. Prevalence is the proportion of the total number of cases (of disease) to the total population rather than the rate of occurrence of new cases. Thus, incidence conveys information about the risk of contracting the disease, whereas prevalence indicates how widespread the disease is. Thus, prevalence is more a measure of the burden of the disease on society.

	Disease	No Disease
Exposure	a	b
No Exposure	c	d

Odds Ratio and Relative Risk: The odds is the ratio of the probability of getting disease over the probability getting no disease. For the exposure and no exposure groups, the odds will be a/b and c/d, respectively (see table above). The odds ratio is defined as $OR = \frac{ad}{bc}$. Relative risk (RR) is a ratio of the probability of the event occurring in the exposed group versus an unexposed group: $RR = \frac{a/(a+b)}{c/(c+d)}$. When a disease is rare, a and c are very small so that $a + b \approx b$ and $c + d \approx d$. Thus the relative risk here is close to the odds ratio.

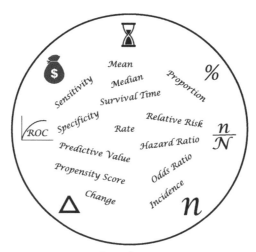

Evidental Measures

Positive and Negative Predictive Values: In diagnostic testing, the positive predictive value is the proportion of positive test results that are true positives (correct diagnoses). Its value depends on the prevalence of the outcome of interest, which may be unknown for a particular target population. The negative predictive value is the proportion of negative test results that are negative, which also depends on the prevalence.

Sensitivity and specificity: Sensitivity measures the proportion of actual positives that are correctly identified as such. Specificity measures the proportion of negatives that are correctly identified. The difference between positive (negative) predicted value and sensitivity (specificity) is that the former uses the overall population as the denominator and the latter uses only the number of true positive (negative) subjects as the denominator.

Hazard Rate and Hazard Ratio: In survival analysis, we often use hazard rate and hazard ratio. Hazard rate or failure rate is the frequency with which patients die or systems fail, expressed, for example, in deaths (failures) per hour. The hazard ratio is the ratio of the hazard rates between two comparison groups.

A *statistic* is a function of measures on a sample. An endpoint can be presented using descriptive statistics, such as mean, median, or mode or using inferential statistics (test statistics) to make directly a statement about population parameters.

An endpoint is usually modeled by a statistical model that represents the causal relationship (or association) between the endpoint and independent factors. A *statistical model* is similar to a mathematical model or equation that relates the outcome (dependent variable) to the factors (independent variables). However, there are differences. In statistical modelling, the outcome (and sometime the factors) are random variables. For example, in cancer studies, we may choose survival as the random dependent variable for the outcome and treatment indication and severity of the disease at baseline as the random variables.

Statistical model provides a convenient and powerful way to model the causal relationship. For example, an exponential model allows us to estimate the parameters or the hazard rates (λ_1 and λ_2) for two treatment groups. The hazard ratio is simply λ_1/λ_2, which is also the ratio of median (mean) survival time between the two groups since median survival time $= \ln(2)/\lambda$ and mean survival time $= 1/\lambda$ under the exponential model.

4.4 Hypothesis Test

There are two distinct problems in probability and statistics: prediction and inference. *Prediction* is to foretell an effect (probability of a sample) when the cause (population parameter) is given, whereas inference, operating in the opposite direction, starts with effect (a known sample) and attempts to infer the cause (the unknown population parameter). In this and the next two sections, we will mainly discuss statistical inference.

Suppose a patient has to decide to see one of two doctors who have experience with his disease: one with a 100% cure rate, having treated and cured only one patient with the disease, and one doctor with a 99% cure rate, who has treated 100 patients, 99 of whom have been cured. We probably believe the second doctor is better since the first doctor may be just lucky in curing the patient, while the second doctor's skill seems pretty reliable. The notion of hypothesis testing is to control the probability of making a false positive claim (e.g., can reliably cure this disease).

A *hypothesis test* usually involves a *null hypothesis* H_o (e.g., "the drug is ineffective") and an *alternative hypothesis* H_a ("the drug is effective"). We usually write the hypothesis we are interested in as the alternative hypothesis so as to control the false positive error or *type-I error* (see later text). For example, if the parameter θ presents the drug effect (the expected response rate in the test group minus the response rate in the placebo group), then the notation "$H_o : \theta \leq 0$" represents the null hypothesis that there is no positive drug effect, and the notation "$H_a : \theta > 0$" represents the alternative hypothesis that there is a positive drug effect. The null hypothesis H_o will be either rejected or not rejected in terms of the so-called p-value that is a function of observed (experimental) data. What we have just discussed is a one-sided hypothesis test. A hypothesis test can be two-sided in the form of $H_o : \theta = 0$ versus $H_a : \theta \neq 0$.

	H_o Is True	H_a Is True
Don't Reject H_o	$1 - \alpha$	Type-II error β
Reject H_o	Type-I error α	$1 - \beta$

The probability of erroneously rejecting H_o when H_o is true is called the false positive rate or type-I error rate, which is controlled at a nominal value α, called the test's *level of significance*. Similarly, the probability of erroneously rejecting H_a when H_a is true is referred to as the *type-II error rate*, which is also controlled at a predetermined level β. Given a value of θ ($> \theta_0$), the probability of correctly rejecting H_o is called the *power* of the hypothesis test.

The *p-value*, ranging from 0 to 1, is a function of the observed data x, measuring the strength of the evidence against H_o. The smaller the p-value is, the stronger the evidence is. It is the probability of getting data that are at least as extreme as x, given that the null hypothesis H_o is true. The p-value will be compared with a nominal threshold α (e.g., 0.05) to determine if the null hypothesis should be rejected. In other words, if the p-value is less or equal to α, we reject H_o; otherwise, we don't reject it.

There are some undesirable properties of hypothesis testing:

(1) The choice of the level of significance α (usually 5%) is somewhat arbitrary.

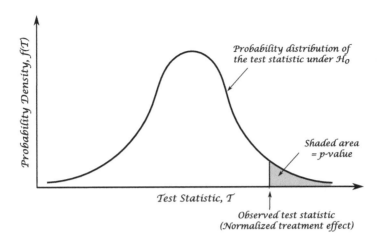

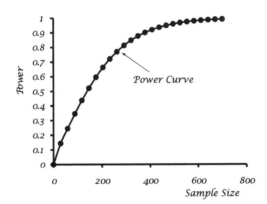

Power Increase as Sample Size Increases

(2) The p-value is only dependent on the tail distribution of the statistic under H_o, but is independent of H_a.

(3) When the sample size is large enough, no matter how small the value of θ is, statistical significance can be achieved.

You may want to ask yourself: why are these undesirable attributes?

It is important to know that the p-value is not the probability that H_o or H_a is true. It is true that the sum of the probabilities that H_o is true and that H_a is true is unity. However, the probability of observing the data when the null is true and the probability of observing the same data when the alternative is true do not necessarily add up to unity.

Sample size is essential in experiment design. When sample size increases the power increases. If the sample size is too small, the power will be too low, meaning the probability of rejecting the null hypothesis is low. If the sample size is too large, it will increase the cost of the experiment. Clearly the sample size and power are not linearly related: after 90%, the power increases slowly as the sample size increases. Power depends on a number of factors: the level of significance α, the magnitude of the parameter θ, the variability of the data σ, and the sample size n. Power increases as the sample size, θ, or α increases and σ decreases. The type of experiment will also often influence the power. For example, with the same sample size, a two-sample parallel design and a 2×2 crossover design will provide different powers.

For complex experimental designs, the sample size, or the power, can be determined using computer simulations. Most recently, a new school of clinical trials has been developed, in which the sample size or other features of the design are not fixed, but can be modified as the observations accumulate, i.e., the adaptive designs discussed in Section 3.10. Simulations are necessary to determine the sample size at various stages of the adaptive design.

4.5   Likelihood Principle

In statistics, a *likelihood function*, or *likelihood* for short, is a function of the parameters of a statistical model. The likelihood of a set of parameter values θ, given outcomes x, is equal to the probability of those observed outcomes given those parameter values, that is,

$$l(\theta|x) = \Pr(X = x|\theta).$$

Mathematically, the likelihood $l(\theta|x)$ is viewed as a function of θ, given x. The likelihood function is often used in estimating a parameter of a statistical model. The value of θ corresponding to the maximum value of $l(\theta|x)$ is called the maximum likelihood estimate of θ.

The *likelihood principle*, an important principle in Bayesian statistics, asserts that the information contained by an observation x about quantity θ is entirely contained in the likelihood function $l(\theta|x)$. Moreover, if x_1 and x_2 are two observations depending on the same parameter θ, such that there exists a constant c satisfying

$$l_1(\theta|x_1) = cl_2(\theta|x_2), \text{ for every } \theta,$$

then they contain the same information about θ and must lead to identical inferences.

Suppose there is a claim against a company for job discrimination against women over the past five years. Let's denote the proportion of men hired in the past 5 years by p. If there is no job discrimination, p should be around

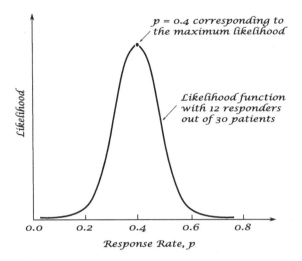

Likelihood of Binomial Distribution with Rate 0.4

0.5. This problem can be formulated as a hypothesis test:

$$H_o : p = 0.5 \text{ vs } H_a : p > 0.5.$$

To test the hypothesis, we need to gather data by randomly selecting employees who were hired in the past 5 years. Assume that no employees left the company in the past five years. The sampling was finished with 3 women out of 12 employees.

Scenario 1: If the total number of employees, $N = 12$, is predetermined, the number of women, X, follows the binomial distribution. The likelihood in this case is given by

$$l_1 = 220 p^9 (1-p)^3.$$

Scenario 2: If the number of women, $n = 3$, is predetermined and the sampling stops as soon as 3 men are selected, then the number of women, X, follows the negative binomial distribution. The (partial) likelihood in this case is given by

$$l_2 = 55 p^9 (1-p)^3.$$

Since the two likelihoods are proportional: $l_1 = 4 l_2$, we can, according to the likelihood principle, conclude that the two scenarios should not lead to different conclusions (rejection or not). However, from a frequentist's perspective these two scenarios can lead two to different decisions because the two test statistics have different distributions under the null hypothesis H_o and thus have different p-values: In scenario 1, the p-value is 0.073. The null hypothesis H_o cannot be rejected at a one-sided level, $\alpha = 0.05$. In scenario 2, the p-value is $0.0327 < 0.05$. Therefore, H_o should be rejected.

Likelihoodists would argue that the fact, "3 women out of 12 people," will not change whether the experimenter's plan is to stop when the sample size reaches 12 or to stop when 3 women are observed. Therefore, the experimenter's intention should not affect the conclusion drawn. Now here's a question for you to contemplate: Would the expected proportion of girls be same if all women in the world decided to stop having children as soon as she had 3 girls or if each woman stopped having children as soon as she reached 12 children?

The likelihood principle is not operational, i.e., it cannot be used to make a decision. Thus, we have to develop some rule for decision making.

When there is no prior information on θ, all values of θ are equally likely before we have the current data. The likelihood tells us how strongly the data support different values of θ. Therefore, it can be viewed as a function of θ, which measures the relative plausibility of different values of θ. This is exactly the notion behind the *Law of Likelihood* (Hacking 1965), which asserts: If one hypothesis, H_o, implies that a random variable X takes the value x with probability $f(x|\theta_o)$, while another hypothesis, H_a, implies that the probability is $f(x|\theta_a)$, then the observation $X = x$ is evidence supporting H_o over H_a if $f(x|\theta_o) > f(x|\theta_a)$.

The hypothesis testing method that arises directly from the law of likelihood is the so-called likelihood ratio test. For a hypothesis test with the null hypothesis $\theta = \theta_0$ against the alternative hypothesis $\theta = \theta_a$, the likelihood ratio is defined as

$$LR = \frac{l(\theta_o|x)}{l(\theta_a|x)}.$$

When the ratio LR is below a predetermined constant threshold c (e.g., 1), we reject the null hypothesis; otherwise, we don't reject it.

4.6   Bayesian Reasoning

The term *Bayesian* refers to the 18th century theologian and mathematician Thomas Bayes. Bayes conceived of and broadly applied a method of estimating an unknown probability on the basis of other, related, known probabilities. Because they infer backwards from observations to parameters, or from effects to causes, the methods of Bayesian reasoning came to be known as inverse probability. The French mathematician Pierre Laplace was the first to introduce a general version of the Bayesian theorem commonly used today (McGrayne, 2012).

Human beings ordinarily acquire knowledge through a sequence of learning events and experiences. We hold perceptions and understandings of certain things based on our prior experiences or prior knowledge. When new facts are observed, we update our perception or knowledge accordingly. No matter whether the newly observed facts are multiple or solitary, it is this progressive, incremental learning mechanism that is the central idea of the Bayesian approach. Bayes' rule (or theorem) therefore enunciates important and fundamental relationships among prior knowledge, new evidence, and updated knowledge (posterior probability). It simply reflects the ordinary human learning mechanism and is part of everyone's personal and professional life (Chang and Boral, 2008).

Suppose our question is whether the investigational drug has efficacy. So far we have finished experiments with the drug on animals, and we are going to design and conduct a small phase 1 clinical trial. Bayesian statisticians would consider the drug effect, θ, a random variable with a distribution. Such a distribution when estimated based on previous experiences or experiments, in this case, animal studies, is called the prior distribution of θ, or simply the prior. This prior knowledge or distribution of the drug effect will be integrated with current experimental data (the clinical trial data), x, to produce the so-called posterior distribution of the drug effect, θ. Generally speaking, the more prior data we have and the more relevant they are, the narrower the prior distribution will be.

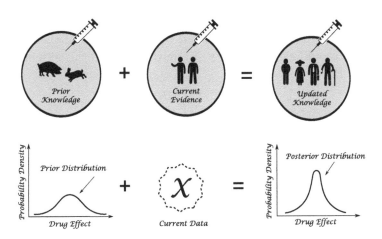

Bayesian Inference: The Bayes Rule

As illustrated, the prior distribution is widely spread, indicating that we have no accurate and precise estimate of the drug effect when there are only animal data available. On the other hand, the posterior distribution is much narrower, indicating that the combination of the animal and human data can provide a much more accurate and precise estimate of the drug effect. Although our knowledge has improved as a result of doing the clinical trial, we still do not know exactly the effectiveness of the drug.

While both frequentist and Bayesian statisticians use prior information, the former use a prior in the design of experiments only, whereas the latter use priors for both experimental designs and statistical inferences (estimations) on the parameter or the drug effect. The frequentist approach considers a parameter to be simply a fixed value, while the Bayesians view that same parameter as a variable with a probability distribution and recognize that the variable is liable to be updated at some time in the future, when yet more information becomes available. One trial's posterior probability is only the next trial's prior probability! Having said that, there is another approach, called Empirical Bayes, in which the prior is calculated from the current data.

The main difference between the Bayesian and frequentist approaches is that Bayesianism emphasizes the importance of information synergy from different sources: the model $f(x|\theta)$, a *prior distribution* of the parameters, $\pi(\theta)$, and current data x. *Bayes' Theorem* places causes (observations) and effects (parameters) on the same conceptual level, since both have probability distributions. However, the data x are usually observable, while the parameter θ is usually latent.

Bayes' Rule plays a fundamental role in Bayesian reasonings. If A and B are two random events, then the joint probability of A and B, due to symmetry, can be written as $\Pr(B|A)P(A) = \Pr(A|B)\Pr(B)$, where $\Pr(X|Y)$ denotes the conditional probability of X given Y. From this formulation, we can obtain Bayes' rule:

$$\Pr(B|A) = \frac{\Pr(A|B)\Pr(B)}{\Pr(A)}.$$

Bayes' rule is useful when $\Pr(A|B)$ is known and $\Pr(B|A)$ is unknown. Here we often call $\Pr(B)$ the prior probability and $\Pr(B|A)$ the posterior probability. The data in the form of conditional probability $\Pr(A|B)$ is the likelihood when it is viewed as a function of the model parameter.

In general, the three commonly used hypothesis testing (H_1 versus H_2) methods in the Bayesian paradigm are

(1) Posterior Probability: Reject H_1 if $\Pr(H_1|D) \leq \alpha_B$, where α_B is a predetermined constant.
(2) Bayes' Factor: Reject H_1 if $BF = \frac{\Pr(D|H_1)}{\Pr(D|H_2)} \leq k_0$, where k_0 is a small value, e.g., 0.1. A small value of BF implies strong evidence in favor of H_2 or against H_1.

(3) Posterior Odds Ratio: Reject H_1 if $\Pr(H_1|D) / \Pr(H_2|D) \leq r_B$, where r_B is a predetermined constant.

Keep in mind that the notions of the Bayes' factor and the likelihood ratio are essentially the same in hypothesis testing, even though there are several versions of likelihood tests.

Practically, we often write Bayes' law in the following form:

$$\Pr(H_1|D) = \frac{\Pr(D|H_1)\Pr(H_1)}{\Pr(D|H_1)\Pr(H_1) + \Pr(D|H_2)\Pr(H_2)},$$

where H_2 is the negation of H_1.

Let's use a court case to illustrate the application of Bayes's rule. Sally Clark was a lawyer who became the victim of an infamous miscarriage of justice when she was wrongly convicted of the murder of two of her sons in 1999. Her first son died suddenly within a few weeks of his birth in 1996. After her second son died in a similar manner, she was arrested in 1998 and tried for the murder of both sons. Pediatrician Professor Sir Roy Meadow testified that the chance of two children from an affluent family suffering sudden infant death syndrome, known as SIDS, was 1 in 73 million, which was arrived at by squaring 1 in 8500, which was given to be the likelihood of a cot death in a family such as Clark's, i.e., stable, affluent, nonsmoking, with a mother more than 26 years old (McGrayne, 2012). The Royal Statistical Society later issued a public statement expressing its concern at the "misuse of statistics in the courts" and arguing that there was "no statistical basis" for Meadow's claim.

A Bayesian analysis would have shown that the children probably died of SIDS as McGrayne showed (McGrayne, 2012):

Let H_1 be the hypothesis to be updated: the children died of SIDS and H_2 is the opposite hypothesis of H_1, i.e., the children are alive or murdered Let data D be that both children die suddenly and unexpectedly.

The chance of one random infant dying from SIDS was about 1 in 1,300 during this period in Britain. The estimated odds of a second SIDS death in the same family was much larger, perhaps 1 in 100, because family members can share a common environmental and/or genetic propensity for SIDS. Therefore, the probability $P(H_1) = \frac{1}{1300} \times \frac{1}{100} = 0.0000077$. Thus, the probability $P(H_2) = 1 - P(H_1) = 0.9999923$. Next, only about 30 children in England, Scotland, and Wales out of 650,000 births annually were known to have been murdered by their mothers. The number of double murders must be much lower, estimated at 10 times less. Therefore, given a pair of siblings neither of whom died of SIDS, the probability that both die a sudden and unexpected death or are murdered is approximately $P(D|H_2) = \frac{30}{650000} \times \frac{1}{10} = 0.0000046$.

And the probability of dying suddenly and expectedly given they die of SIDS is $P(D|H_1) = 1$.

The goal is to estimate $P(H_1|D)$, the probability that the cause of death was SIDS, given their sudden and unexpected deaths. Bayes' rule provides the formula

$$P(H_1|D) = \frac{1 \times 0.0000077}{1 \times 0.0000077 + 0.9999923 \times 0.0000046} = 0.626$$

Thus, it is in fact likely that the infants died of SIDS, definitely not a 1 in 73 million chance.

The convictions were upheld at appeal in October 2000 but overturned in a second appeal in January 2003. On the day she won her freedom, Sally issued a statement in which she said: "Today is not a victory. We are not victorious. There are no winners here. We have all lost out. We simply feel relief that our nightmare is finally at an end. We are now back in the position we should have been in all along and plead that we may now be allowed some privacy to grieve for our little boys in peace and try to make sense of what has happened to us."

Following the case, another three wrongly convicted mothers were declared innocent of crimes that had never occurred. However, Sally never recovered from the experience and died in 2007 from alcohol poisoning. Clark's case is one of the great miscarriages of justice in modern British legal history.

There are recent dramatic developments in the applications of the Bayesian approach, thanks to efficient computer algorithms such as *Markov chain Monte Carlo* techniques (MCMC) and new computer simulation algorithms found to be useful in determining the complex Bayesian posterior distribution, an outcome which was previously considered impossible.

4.7   Causal Space

In a broader sense, scientific discovery is about identification of similarities, and these similarities can be in the sense of either "cause" or "effect." The concept of *causal space* introduced here refers to the collection of similar conditions which provide a vantage point, from which researchers' causal reasoning takes place. It is a way of grouping similar things in order to find scientific laws. For instance, we consider all smokers are similar (regardless how much and when they smoke), putting them into one group and nonsmokers into another, no matter how different they are in many aspects, so that we can have a "scientific law": A smoker is more likely to develop lung cancer than a nonsmoker. On the other hand, we may classify subject behaviors into heavy drinking, moderate drinking, and no-drinking categories so that we can "discover" the "law" that moderate drinking is good for the heart. The consequence of the grouping is a causal space containing smoking and moderate drinking. Such grouping is based on similarity and thus is subjective. The notion behind the grouping is the similarity principle.

Because of subjectivity in choosing a causal space, there can be many controversial scientific conclusions. Suppose a Chinese man was taking a trip from China to the United States. When he crossed the border of the U.S. he wondered if his life expectancy should be calculated according to U.S. population or Chinese population. For the U.S. population the life-expectancy is 78.6 years and for Chinese it is 75.6 years (WHO, 2011). If he should not use the U.S. life expectancy at the moment, when can he, 5 or 10 years later? Clearly, how to formulate the similarity set or the causal space will directly affect the conclusion we are going to draw. In fact, everyone's life span is fixed but unknown until he dies. Thus, we use the average life expectancy of a set of similar people (the causal space) to infer his expectancy.

The causal space is different in principle for different statistical paradigms. For hypothesis testing in the frequentist paradigm, the causal space depends on the experimental procedure. For instance, the causal spaces will be different for an experiment with a fixed sample size and a sequential experiment with optional stopping. In the example we discussed in the Likelihood Principle section, all experiments with a preset value $N = 12$ (total number of games) will be considered as a group in the causal space. The number of women, X, follows the binomial distribution. Similarly, all experiments with prefixed $n = 3$ will be considered as another group in the causal space. The number of women, X, follows the negative binomial distribution.

In contrast, a Bayesian's and likelihoodist's causal space does not depend on the experimental procedure. For instance, the causal space will be the same for an experiment with a fixed sample size or with sequential stopping. For the previous example, all experiments with $N = 12$ and $n = 3$ will be considered belonging to a group in the causal space, i.e., it doesn't matter whether the N or n is prechosen.

Different statistical paradigms use different causal spaces for inferences and predictions. To judge which is better, we have to evaluate its "repetitive performance," not a single-time performance, and must propose a criterion for the evaluation. Such a criterion can be subjective. For example, one frequentist criterion is the type-I error in hypothesis testing, while a Bayesian criterion would be the accuracy of a posterior distribution.

We often hear from statisticians that drug effects from an animal experiment and drug effects from a human experiment (clinical trial) are independent, because these are two different (independent) experiments and are statistically independent. On the other hand, animal data and human data are related; otherwise why do we test a drug on animals before we test it on humans? This question can be answered clearly in terms of relationships at individual and aggregative levels and in terms of the general concept of *causal space*.

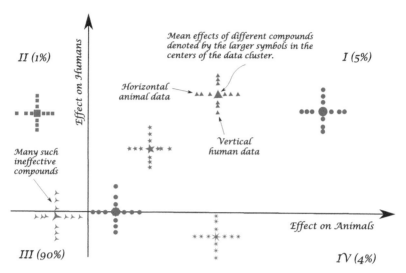

Hypothetical Drug Effects on Animals and Humans

We believe that there are many chemical compounds (maybe 90% of all compounds undergoing animal studies) that are ineffective in treating disease for both animals and humans. Some compounds are effective in animals but not in humans. Only a small portion of compounds are effective in both animals and humans, but their magnitudes of effectiveness are generally different. Such a belief is illustrated in the figure above, where the mean effects on an animal population and a human population are denoted by the larger geometric symbols (square, circle, triangle, star, etc.), vertically surrounded by individual human data (the smaller symbols) and horizontally surrounded by individual animal data (the smaller symbols).

It is extremely important to differentiate between two types of relationships: the individual level and the aggregative (mean) level. At an aggregative level, the relationship between the mean responses of humans and the mean responses of animals to different compounds (the larger symbols in the figure) is not trivial and can be characterized by the percentages in the four quadrants: hypothetically, 5%, 1%, 90%, and 4% for Quadrants I through IV, respectively). However, the individual responses of animals (the small horizontally aligned symbols) and the individual responses of humans (the small vertically aligned symbols) for a given compound are trivial or independent.

Our ultimate goal is to discover an effective compound for humans and estimate the magnitude of its effectiveness. Since the majority of compounds are ineffective, a stepwise approach is appreciated. First we identify the

compounds on the right side of the vertical axis. To this end, we formulate the causal space by grouping all compounds on the left side (ineffective
compounds on animals) as one category and those on the right side (effective
compounds on animals) as the other category. The percentages of these two
categories in this hypothetical example are 91% and 9%, respectively. In this
step we can remove 90% of the negative compounds; for the moment, we ignore the potential false positive and false negative errors. Even though we may
miss some (1%) interesting compounds in Quadrant II, this approach is effective because animal experiments are relatively low cost and safe to humans in
comparison to experiments in humans (clinical trials). In this step, we have
used the relationship between animals and humans at the aggregative level,
in other words, the mean responses to compounds. The next step is to further
differentiate the compounds in Quadrants I from those in IV. To do this we
have to run clinical trials. The causal space consists of two groups: compounds
in Quadrants I and IV, respectively. The percentages of compounds (5% and
4%) in these two quadrants characterize the relationship between animals and
humans at the aggregative level.

To determine the magnitude of the effectiveness of a drug, a frequentist statistician will use only the data from human experiments of the test
compound, whereas a Bayesian statistician will combine the data from both
animal and human studies. Frequentists consider that all humans are similar
in response to the test compound and group them together to form the causal
space, whereas Baysianists have a great flexibility in forming the causal space:
for example, Baysianists may consider that humans are similar in response to
a test compound but animals are also similar (but less similar) to humans.
Therefore, animals and humans can be grouped together to form the causal
space. The difference in similarity (human–human versus animal–human) is
considered when using animal data to construct the prior distribution of the
treatment effect on humans. In the Bayesian paradigm, the causal space can
also be formulated on the basis of similarity between human responses to different drugs. As a result, we can construct the prior using previous human
responses to different drugs, which will then be combined with the results
from the clinical trials for the current test drug to calculate the treatment
effect. There are many other possible ways to define similarity in forming the
causal space because two things are always different and similar at the same
time. Similarity is always in a relative sense: an animal may be less similar
to a human than a human, but more similar to a human than a stone. The
bottom line is that there is no unique definition of similarity when applying
the similarity principle. The final conclusion will be different if we explicitly
or implicitly use different similarity definitions or causal spaces. The higher
the similarity is and the more data that are available, the more precise the
conclusion will be. The problem is that there is always a tradeoff between

similarity and the amount of data available. We can call for either a high similarity with less data or more data with a low similarity. It is generally true that the Bayesian approach emphasizes more data and that the frequentist approach stresses a high similarity.

We should remember from Section 3.2 that in drug development, before an in-vivo (animal) study is carried out, we have to conduct in-vitro (isolated issues) experiments, for reasons of cost effectiveness and animal protection.

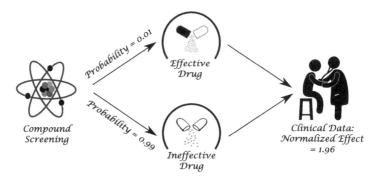

Prior Information in Drug Development

We now discuss how we reach conclusion form different statistical paradigms. We realize that the effect of a test drug or chemical compound on a population is fixed but unknown until it is fully tested on the entire population. But how is it that we infer the effect of a drug on an entire population when it has been tested in only a sample of the population? For simplicity, let's assume the true effects, θ, of effective and ineffective drugs are 1 and 0, respectively. The test drug is selected from a chemical compound library, stored with a mixture of effective and ineffective compounds, and then is tested on a sample of patients—the clinical trial (I oversimply the process for the moment). Presumably, the probability of selecting an ineffective drug is 99% and an effective compound is 1%, and we observed that the normalized effect of the test drug is $x = 1.96$ in the clinical trial.

The question is: Is the test drug effective?

For frequentists, the causal space is the target patient population. Since $x = 1.96$ from the clinical trial, the corresponding p-value is 0.025, no larger than the one-sided significance level 0.025. Therefore, we reject the null hypothesis that the test drug has no effect and conclude the drug is effective. For likelihoodists, the causal space is also the target patient population. Since the likelihood ratio is $LR = \frac{l(\theta=0|x=1.96)}{l(\theta=1|x=1.96)} = \frac{0.0584}{0.2516} = 0.232$ is much smaller than 1, we hold in favor of $\theta = 1$.

For Bayesian statisticians, the prior knowledge about the probability of selecting an effective compound is also of importance. However, since the selecting probability is unknown, a "subjective" probability of selecting an ineffective compound must be used here. After combining the prior (assume $\Pr(\theta = 0) = 0.99$, $\Pr(\theta = 1) = 0.01$) with the data from the clinical trial using Bayes' rule, the posterior probability is $\Pr(\theta = 0|x) = \frac{0.0584(0.99)}{0.0584(0.99)+0.2516(0.01)} = 0.958$, and we conclude the drug is very likely an ineffective one.

In reality, since drug development requires several stages (compound screening, an in vitro test, an in vivo test, preclinical experiments, and Phase-1 to Phase-3 clinical trials), at Phase-3 stage approximately 50% of test drugs are likely to be effective. This estimate is based on the fact that about 40% of test drugs in Phase-3 are finally approved for marketing, and Phase-3 trials are usually implemented with 90% power. If we are using the prior $\Pr(\theta = 0) = 0.5$, the posterior $\Pr(\theta = 0|x) = 0.188$ and so we conclude the drug is likely an effective one. Bayesian approach can be sensitive to the prior; therefore, it is difficult to use in the final stage of the drug approval process.

4.8   Decision Theory

We are making decisions, big or small, every waking minute of our daily lives and, for scientists, in the process of doing research. There are two distinctive approaches to decision-making as delineated by descriptive and normative decision theories. *Descriptive decision-making theory* considers a decision as a specific information processing process; it is a study of the cognitive processes that lead to decisions, for example, the ways humans deal with conflicts and perceive the values of the solutions. Descriptive decision making seeks explanations for the ways individuals or groups of individuals arrive at decisions so that methods can be developed for influencing and guiding the decision process. *Normative decision-making* considers a decision as a rational act of choosing one or more viable alternatives. It is a mathematical or statistical theory for modeling decision-making processes. Frequentist (classical) decision-making theory and Bayesian decision theory fall into this category. Normative decision making strives to make the optimal decision, given the available information. Hence, it is an optimization technique.

The normative decision-making approach can be further divided into deterministic and probabilistic approaches. For the deterministic approach, information is considered as completely known, whereas for the probabilistic approach, information is associated with uncertainties or probabilities.

In decision theory, statistical; models involve three spaces: the observation space, which consist of all data; the parameter space, which is the collection of all possible situations; and the action space A, which is the compilation of all possible actions. Actions are guided by a decision rule. An action always has an associated consequence that is characterized by the loss (or utility) function L, which usually is a function of the action taken and the situation under consideration.

Statistical analyses and predictions are motivated by objectives. When we prefer one model to an alternative, we evaluate potential consequences or losses. However, different people have different perspectives on the losses and, hence, use different loss functions. Loss functions are often vague and often not explicitly defined, especially when they pertain to decisions we make in our daily lives. Decision theory makes this loss explicit and deals with it with mathematical rigor.

Because it is usually impossible to uniformly minimize the loss in all the possible scenarios, a decision rule is determined so that the average loss is minimized. Frequentist hypothesis testing is based on the notion of the type-I error rate (which actually is the worst-case scenario approach) by assuming H_o does happen and by controlling the hypothetical error rate. On the other hand, hypothesis testing can be explained from a decision perspective: without the false positive error control, researchers will behave differently, e.g., intentionally make the result better. Such behavioral changes should be considered in the loss function, while setting a small value for the level of significance α in the hypothesis test can be loosely viewed as a loss control in decision theory.

In *Bayesian decision theory*, we consider a set of possible scenarios, the probability of each, and the impact (loss) if a scenario actually materializes. The probability of a potential event coming to pass can be the posterior probability, calculated from prior and current data. An action that minimizes the posterior expected loss is called a Bayes' action.

Decision theory not only can involve probabilities, but also can involve a chain of hierarchical decisions. Here is a common sequential decision problem.

A typical *sequential decision problem* concerns a dynamic system or a process that is characterized by a sequence of states $A, B, C, \dots$. The process can be (probabilistically) controlled by choosing a sequence of actions $a, b, c, \dots$ at different time points. For each action taken, there are an associated reward (cost) and transition probability from one state to the next. The goal is to find a policy or strategy (a sequence of actions/decisions) that maximizes the expected reward.

There are two common ways to solve a decision problem: Forward induction and backward induction. A forward induction starts from time 0 and works forward to the end, while backward induction starts from the end and works backward to the beginning. Forward induction is intuitive and works well for simple problems, whereas backward induction is more advanced and can be used for more complicated problems.

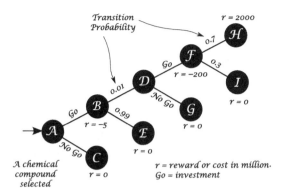

Decision Tree

Suppose a promising compound is selected after an initial screening. The compound may be effective and can potentially be developed into a drug. Drug development can be very costly but at the same time very profitable if the drug is eventually approved for marketing. Here is a sequence of decisions to be made (see above diagram): At state A, if we decide not to develop the drug (the no-go decision), there will be no cost and no gain, while if we decide to develop the drug (a go decision), there will be a cost (negative reward) of $5 million. At state B, we know that the drug has a 1% probability of being effective and 99% probability of being ineffective. If it is ineffective, we will get zero reward; if it is effective (state D), we have another decision to make: if we make the no-go decision, there will be no reward; if we make the go decision, the cost will be $200 million. At state F, there are one of the two possible outcomes: (1) the drug fails to show effectiveness (with a probability of 30%) or (2) the drug demonstrates effectiveness with an award of $2,000 million. For the forward induction, we calculate the expected rewards for all three possible sequences of actions: (1) no-go. (2) go then go, and (3) go then no-go. The expected reward is equal to the sum of reward × probability products on the path, which are 0, 7, and −5 million, for the three decision sequences, respectively. Therefore, the strategy: {go, go} is the best policy with a reward of 7 million. These expected rewards are calculated at state A. When state D reached, the rewards should be recalculated.

The *backward induction* is based on the notion of Bellman's optimality principle (1957). This principle states that an optimal policy has the property that, whatever the initial state and initial decision are, the remaining decisions must constitute an optimal policy with regard to the state resulting from the first decision. An application of Bellman backward induction in project scheduling will be discussed in Section 7.7.

4.9   Statistical Modeling

A *statistical model* is a quantitative characterization of relationships between deterministic and random variables. Statistical models are different from mathematical models in that the former involve random variables that are stochastically, not deterministically, related. A statistical model allows us to make causal inferences and predictions. To illustrate this, suppose that in the experiment of the Tower of Pisa, Galileo dropped a cannon ball from various heights (H) of the tower and recorded the times $(\hat{T})$ it took for the ball to fall.

Data from Cannon Ball Falling Experiment

H (meter)	20	30	40	50
$\hat{T}$ (second)	2.1	2.3	2.8	3.3
$.45\sqrt{H}$ (second)	2.0	2.5	2.8	3.2

We use our knowledge from physics and propose the following model.

$$T = c\sqrt{H},$$

which means that when the height increases four times, the time to fall increases only twice. From physics we know that $g = 2/c^2$ is Earth's gravitational acceleration.

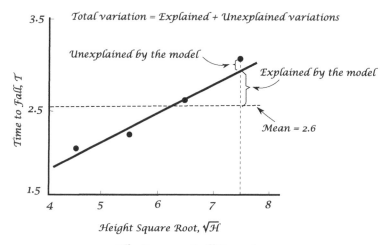

The Cannon Ball Experiment

There are random errors such as measurement error and the natural variation of g from different locations and heights at which the experiments take place. To determine the constant c, we need to define a measurement of overall error and optimize the model by minimizing the overall error. However, such a definition of error and optimization are not unique. Suppose we define the error to be the *mean square error* (MSE) between the predicted time T and the observed time $\hat{T}$. For instance, for $c = 0.45$, the mean squared error is

$$MSE = \frac{(2.1 - 2.0)^2 + (2.3 - 2.5)^2 + (2.8 - 2.8)^2 + (3.3 - 3.2)^2}{4} = 0.015.$$

We try different values of c until we get the minimal MSE. This is called the *mean square error method*.

However, if we define the error as the *mean absolute difference* (MAD) between T and $\hat{T}$, then we will have a different c that minimizes the error (MAD).

$$MAD = \frac{|2.1 - 2.0| + |2.3 - 2.5| + |2.8 - 2.8| + |3.3 - 3.2|}{4} = 0.1$$

The mean square error method is often used mainly because of (unfortunately!) the mathematical simplicity. Of course, there are better methods than the trial–error method to obtain the model coefficient c, as found in standard statistical books and statistical software packages.

After we find the optimal model, an immediate question might be: How good is the model in fitting the data, and can it be used for prediction? Statistically the question has two components: (1) How close are the predicted values to the observed data values? (2) Does the model fit the data accidentally or is the model statistically significant? The first question can be answered statistically using the so-called R^2 statistic, which usually ranges from 0 to 1. The mean model, which uses the mean for every predicted value, generally would be used if there were no informative predictor variables. The R^2 value is 0 for the mean model. In the cannon ball experiment, the mean (average) time is 2.6. Therefore, the fit of a proposed regression model should be better than the fit of the mean model. R^2 is often explained, though not very precisely, as the proportion of variability that can be explained by the model. Remember that R^2 is not any real number squared. If one uses a model worse than the mean model (never do that!), then R^2 can be negative. Practically speaking, a value of 0.8 or higher is often considered good for prediction.

We now know that we can often make use of R^2 to measure the goodness of model fit. Unfortunately things are not that simple. Remember that in Section 1.5 we discussed the parsimony principle and concluded that we can construct a model that goes through each data point we have already collected. But such a "perfect fitting" with $R^2 = 1$ is likely accidental and we will not have the same luck next time to see the same model again perfectly fitting a new set of data. To check if model fitting is accidental, we can use a hypothesis test such as the so-called the F-test. The F-test evaluates the null hypothesis that R^2 is smaller than or equal to zero, or equivalently the null hypothesis that all coefficients (only one in the current case) are equal to zero versus the alternative that at least one is not. The F-test determines whether the proposed relationship between the response variable and the set of predictors is statistically reliable, and can be useful when the research objective is either prediction or explanation.

In addition to the F-test, we sometimes use an adjusted R^2 instead of R^2 to measure the goodness of model fit. As we have elucidated, R^2 can increase as factors are added to the (regression) model, even if the factors are not real predictors. To remedy this, a related statistic, the adjusted R^2, incorporates a penalty on the R^2 when adding more factors. Similar to R^2, the adjusted R^2 can be interpreted as the proportion of total variance that is explained by the model.

In reality, statistical models often involve more than one *predictor*. For instance, to model the drug effect we may use three different predictors, a treatment indicator T (0 for a placebo and 1 for the test drug), gender Z (0 for female, 1 for male), and age X (can be considered as a continuous variable). We use Y to represent the dependent variable or clinical efficacy endpoint. Thus, the linear model (linearity is really only a restriction on the parameters/coefficients not on the variables) will be something like this:

$$Y = Y_0 + aT + bZ + cX,$$

where Y_0, a, b, and c are constants (model parameters) to be determined.

For a woman who takes placebo, the expected response is $Y = Y_0 + cX$, where c is the response increase per age unit increase. Similarly, for a man who takes the test drug, the expected response is $Y = Y_0 + T + Z + cX$.

What we have discussed so far about the statistical model is purely from the frequentist's perspective. In the frequentist paradigm the model parameters are considered fixed values. In contrast, in Bayesian statistical models, the model parameters (coefficients) are considered random and have prior distributions. The posterior distributions of the parameters that incorporate data are presented with the updated model. In other words, the uncertainties of the parameters Y_0, a, b, and c are reflected in the posterior distributions that are constantly updated as data accumulate.

4.10 Data Mining

Both *data mining* and *machine learning*, lying at the confluence of multiple disciplines, including statistics, pattern recognition, database systems, information retrieval, the World Wide Web, visualization, and many application domains, have made great progress in the past decade. Data mining usually involves streamed data, which are usually of vast volume, change constantly, and contain multidimensional features. Statistics plays a vital role in data mining.

Learning algorithms in data mining fall into three categories, supervised, unsupervised, and reinforcement learning, based on the type of feedback that the learner can get. In supervised learning, the learner (statistical model) will give a response based on an input and will be able to compare his response to the target (correct) response. In other words, the "student" presents an answer for each input in the training sample, and the supervisor provides the correct answer and/or an error message associated with the student's answer. Based on the feedback, the learner modifies his behavior to improve future performance.

Supervised learning (classification) has been used in digital imaging recognition, signal detection, and pharmacovigilance. It has also been used in the diagnosis of disease (such as cancer) when more accurate procedures may be too invasive or expensive. In such a case, a cheaper diagnosis tool is often helpful to identify a smaller set of people who likely have the disease for further, more advanced procedures. Typically parameters in the model for predicting the disease are determined via the so-called training data set, in which the true disease status and diagnosis results are known for each individual. Supervised learning can also be nonparametric, for instance, the nearest-neighbor method. The notion of this method is that each person has compared certain characteristics with his neighbors (e.g., those in the same city or whatever is defined) who have or don't have the disease. If the person has the characteristics, to a close or similar degree, to those among the group who have the disease, we predict the person will also have the disease; otherwise, the person is predicted to not have this disease.

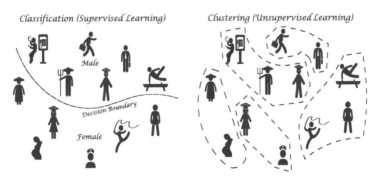

Supervised and Unsupervised Learnings

In *unsupervised learning*, the learner receives no feedback from the supervisor at all. Instead, the learner's task is to re-represent the inputs in a more efficient way, as clusters or with a reduced set of dimensions. Unsupervised learning is based on the similarities and differences among input patterns. The goal is to find hidden structure in unlabeled data without the help of a supervisor or teacher providing a correct answer or degree of error for each observation.

A typical example of unsupervised learning is a self-organizing map (SOM) in data visualization. SOM is a type of artificial neural network (ANN) that is trained using unsupervised learning to produce a low-dimensional, discretized representation of the input space of the training samples, called a map. SOM forms a semantic map where similar samples are mapped close together and dissimilar ones apart. The goal of learning in the self-organizing map is to cause different parts of the network to respond similarly to certain input patterns. This is partly motivated by how visual, auditory, or other sensory information is handled in separate parts of the cerebral cortex in the human brain

Reinforcement learning (RL) is an active area in artificial intelligence study or machine learning that concerns how a learner should take actions in an environment so as to maximize some notion of long-term reward. Reinforcement learning algorithms attempt to find a policy (or a set of action rules) that maps states of the world to the actions the learner should take in those states. In economics and game theory, reinforcement learning is considered a rational interpretation of how equilibrium may arise. A stochastic decision process (Section 5.9) is considered a reinforcement learning model.

RL is widely studied in the field of robotics. Unlike in supervised learning, in reinforcement learning, the correct input–output pairs are never presented. Furthermore, there is a focus on on-line performance, which involves finding a balance between exploration of uncharted territory and exploitation of one's current knowledge. In doing so, the agent is exploiting what it knows to receive a reward. On the other hand, trying other possibilities may produce a better reward, so exploring is sometimes the better tactic.

An example of RL would be game playing. It is difficult to determine the best move among all possible moves in the game of chess because the number of possible moves is so large that it exceeds the computational capability available today. Using RL we can cut out the need to manually specify the learning rules; agents learn simply by playing the game. Agents can be trained by playing against other human players or even other RL agents. More interesting applications are to solve control problems such as elevator scheduling or floor cleaning with a robot cleaner, in which it is not obvious what strategies would provide the most effective and/or timely service. For such problems,

RL agents can be left to learn in a simulated environment where eventually they will come up with good controlling policies. An advantage of using RL for control problems is that an agent can be retrained easily to adapt to environmental changes and can be trained continuously while the system is online, improving performance all the time.

4.11   Misconceptions and Pitfalls in Statistics

There are many misconceptions and misuses of statistics. In this section we will discuss a few such errors, especially those that can be rectified by a proper understanding of fundamental concepts.

(1) *Statistical Significance and Practical Significance*

We often think that achieving statistical significance means success, and thus that the power of a hypothesis test is merely the probability of success renamed. But this is usually not the case. First, power is the probability of rejecting the null hypothesis only when the assumed parameter value is true. In reality, the parameter value is an unknown; otherwise, we wouldn't need to carry out an experiment. Second, there is usually another requirement, called practical significance. We can sometimes find a statistically significant difference that has no discernible effect in the real world.

For instance, suppose the factory that makes Pumpkin Flax Granola Cereal fills the cereal boxes using an automated filling system. Such an automated system cannot be exact, and presumably the actual mean filling weight is 499.8 grams, a gram off the targeted 500 grams. The small difference of 0.2 grams can be shown statistically significant with a sufficiently large sample size, but the difference will not be of practical significance: no one would care about such a tiny variation.

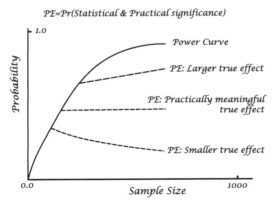

Power and Probability of Effectiveness (PE)

In recent years as the patents of top drugs expire, many companies have started to develop biosimilar drugs, which are generic versions of the brand drug with minor differences in chemical or protein structure. To prove that a generic drug has a similar effect as the brand drug, a noninferiority clinical trial is required. What one tries to prove in a noninferiority trial is that the generic drug is not worse than the original brand drug by a so-called noninferiority margin, which must be no more than the minimal clinically important difference. A drug that prolongs survival time by one week may not be considered practically important because of the possible side effects of the drug, the inconvenience of administrating drug, and the cost in comparison with other available options. We see that a practically important difference is a combination of many different aspects, except the statistical aspect. On the other hand, a difference that appears to be small does not necessarily lack practical meaning. Suppose a pervasive medical condition is treatable with two competing drugs. One has a cure rate of 70% and the other has a 70.1% cure rate. The 0.1% difference seems not to be practically significant. However, if each year there are 1,000,000 new incidences of the condition in the world, 10,000 more patients cured every year is a quite meaningful thing.

The determination of the minimum clinical or practical meaningful difference is difficult and subjective. In most cases there is no clear cut dividing line. For example, in treating hypertension, physicians may say that any reduction in blood pressure is helpful in reducing the risk of heart attack. Well, how about statistical significance? Is there a clear cut separation based on p-value? The answer is: No, the choice of 5% for alpha is subjective; there is nothing magical about 5%.

(2) *Hypothesis Testing and p-Value*

Should one use a one-sided or two-sided test and alpha?

Hypothesis tests can be one-sided or two-sided. For a two-sided test, the null hypothesis concerning a parameter δ (e.g., the mean treatment effect of a test drug) is often written as $H_o : \delta = 0$. Realistically, it is virtually impossible that δ will be exactly equal to zero, especially when test an effective of a chemical compound. Therefore, as long as the sample size is large enough, we don't even need an experiment—we know that H_o will be rejected. Whether to use a one-sided or two-sided test should be determined by the objective of the study. For example, if we are interested in the effectiveness of a test drug, u_t, in comparison with a control drug's effectiveness, u_c, then the hypothesis can be written as

$$H_o : u_t <= u_c \text{ versus } H_a : u_t > u_c.$$

A two-sided test is traditionally used in drug development, but the regulatory authority would never approve a worse drug unless it could provide other benefits. Therefore, there is no reason to combine the error of making the false positive claim that the test drug is better than the control and the error of making the false negative claim that the test drug is worse than control into a single error. Even though using a two-sided test with a level of significance α is numerically equivalent, in a simple experiment, to using a one-sided test at level of significance $\alpha/2$, these are not equivalent in a complicated design such as an adaptive design.

For simplicity, we often write the above hypothesis test for the drug in the form of the so-called point test:

$$H_o : u_t - u_c = 0 \text{ versus } H_a : u_t - u_c = 0.3,$$

where 0.3 (or any other nonzero value) is an example of the estimated treatment difference for the purpose of sample size estimation. Assume next that the one-sided alpha is 0.025 and the observations of treatment effect on individual patients are from a normal distribution with known standard deviation. Here are some verbal habits (traps) that a harried researcher might too quickly fall into:

Misconception 1: "When H_o is rejected, we conclude the treatment different is 0.3." This is incorrect. When H_o is rejected we only accept the negation of H_o, i.e., $u_t - u_c > 0$. The treatment difference $u_t - u_c = 0.3$ in the alternative hypothesis is used only for the purpose of sample size calculation.

Misconception 2: "If H_o is rejected, the observed treatment difference, e.g., $u_t - u_c = 2.8$, is statistically significant; therefore the difference is the true difference." No, this is not true either. We reject H_o because it is very unlikely we can observe a mean difference equal to or larger than 2.8 when H_o is true. The confidence interval and Bayesian credible interval are more informative about true treatment difference.

Misconception 3: "In order to reject H_o in the above hypothesis test, the observed treatment difference must be at least 0.3." No, again incorrect. When the experiment is designed with 90% power, an observed treatment difference of 0.182 will result in a p-value of 0.025, leading to the rejection of the null hypothesis H_o; if an 80% power is chosen, a 0.21 observed difference in treatment will generate a p-value of 0.025 and lead to rejection of H_o. However, given a 50% power, a treatment difference of 0.3 or more is required to reject H_o. Therefore, if we design an experiment with 90% power based on the minimal clinically important difference, say, 0.3, then a 0.181 observed difference with no clinical significance can be statistically significant. Now here is a question for you: Would this design be appropriate or over-powered?

You may wonder why we don't test to see if a drug's effect is larger than the minimal practical difference? In fact, this might be a better approach if we could agree on the value of the minimal practical difference before testing the hypothesis.

(3) *Confidence Intervals and Credible Intervals*

In frequentist statistics, the confidence interval (CI) is commonly reported as an outcome of clinical trials in addition to *p*-values. Ironically, frequentist CIs are often misinterpreted as Bayesian credible intervals (BCIs), though they are often numerically close.

The concept of a CI may be difficult for nonstatisticians. As an example, assume that our parameter of interest is the population mean. What is the meaning of a 95% CI in this situation? The correct interpretation is based on repeated sampling. If samples are drawn repeatedly from a population, and a CI is calculated for each, then 95% of those CIs should contain the population mean. We can say that the probability of the true mean falling within this set of CIs with various lower and upper bounds is 95%. However, we cannot say that the probability of the true mean falling within a particular sample CI is 95% because the population mean and the CI from a particular sample are fixed, and so no randomness is involved here.

The BCI is easy to interpret. If, for example, the 95% BCI of the treatment mean is (1.0, 3.0), we are saying that, based on the data, there is a 95% probability that the true treatment mean falls between 0.1 and 3.0.

A one-sided 95% confidence interval for a parameter u and rejection of the one-sided hypothesis $H_o : u \leq 0$ at a significance level of 5% have the following simple relationship: rejection of H_o is equivalent to the confidence lower limit excluding zero. However, a common mistake is to say that if the two confidence intervals for parameters u_1 and u_2 overlap, then u_1 and u_2 are not significantly different. This is completely wrong. One can easily prove it is wrong by computing the individual CIs for u_1, u_2 and $u_2 - u_1$.

(4) *Level of Significance (α) and False Discovery Rate (FDR)*

A level of significance of alpha $= 5\%$ does not mean there are 5% false positive findings. This is because if all the null hypotheses under investigation are true, then all positive findings are false, no matter what value of α is used. In contrast, if all null hypotheses we investigate are in fact false, there will be zero false findings, independent of α. If some proportion R of all null hypotheses are true, the maximum proportion of false positive findings among all findings is expectedly

$$FDR == \frac{\alpha \times R}{\alpha \times R + \text{power} \times (1 - R)}.$$

Because power is not constant, the above equation can be understood as an average of the ratios. Thus, reducing R or increasing the power can reduce false positive findings or, more accurately, reduce the FDR.

Of course, there are many other nonhypothesis-based findings that have not been considered here.

One may wonder: since different errors have different impacts on our lives, why should we control error instead of the impact of error. For example, a false positive error has a much greater impact when testing a drug against a placebo than when testing the drug against the best drug available in the market. Ironically, the same α is used in the current regulatory setting. I guess it is for the convenience of regulatory authorities in making decisions, a possible practical reason.

(5) *Missing Data Not the Same as No Information*

You may think that when pieces of data are missing we cannot do anything and should just ignore them. However, missing does not equal no information. Instead, missing data can sometimes inform us about something important. In clinical trials, if one clinical center has much more missing data than other clinical centers, it might indicate a problem in trial conduct and monitoring. Larger missing data due to a lack of efficacy can greatly impact the results of analysis if the missing data are simply ignored. Another example would be if a broken instrument cannot read very high and very low values, so that only middle values are recorded. If group A has more high values and group B has more low values, then analyses without missing values will bias the result. There are many ways to deal with different types of missing data (Chang, 2011).

(6) *Too Many Tests and Overfitting*

In statistical analysis, we often try many different models and different hypotheses. Every time we try different models or hypotheses we potentially make false claims. Therefore, the more models or hypotheses we have tried, the more false positive errors we could possibly make. This phenomenon is called the multiplicity issue in statistics. It is probably the most highly impactful and challenging statistical and scientific problem for which we do not have good solutions yet. We will further discuss this issue at the end of Chapter 6.

Misconception 4: "As long as I preset statistical criteria for selecting models or factors in a model and let a computer program automatically determine the model for me (such as the stepwise elimination method in regression), I need not worry about multiplicity." This too is incorrect. By doing auto-fitting to pick the model with the best fitness, one actually commits a kind of overfitting and inflates a type-I error rate.

Chapter 5

Dynamics of Science

5.1 Science as Art

Broadly speaking, science includes *natural science* (physics, chemistry, life science), *engineering science, computer science, mathematics*, as well as the *social sciences, science education*, and other branches.

According to the American Association for the Advancement of Science (AAAS), science must be considered as one of the liberal arts and should be taught with this in mind. To teach Boyle's Law without reflection on what *law* means in science, without considering what constitutes evidence for a law in science, and without attention to who Boyle was, when he lived, and what he did, is to teach in a truncated way (Gauch, 2003; Matthews, 1994).

The utilization of science and the education of science involve humans and human subjectivity. A good example is in drug development, in which the processes of experiment design, conducting the experiment, getting drug approval for marketing, and even the doctor's prescription, all involve the nature of art.

In our education systems we often emphasize the absolute truthfulness and objectivity of science and overlook its inexactness and subjective aspects. Ironically, such a practice itself is not objective or scientific. Indeed, proximity is viewed as an essential characteristic of science, while such proximities to the truth is often subjective. A science teacher may reward his young student for the answer: "A flamingo has 3 legs" because "3 legs" is closer to the truth of 2 legs, or he may reward another student instead for her answer, "A flamingo has 4 legs" since 4 is closer than 3 to 2 because both 4 and 2 are both even numbers.

Science is making progress daily. A scientific revolution occurs when scientists encounter anomalies that cannot be explained by the universally accepted paradigm within which scientific progress has previously been made. The paradigm, in Kuhn's view (1977), is not simply the current theory, but the entire worldview in which it exists, and all of the implications which come with it. It is based on features of the landscape of knowledge that scientists can identify around them. There are anomalies for all paradigms that are brushed away as acceptable levels of error, or simply ignored and not dealt with. Rather, anomalies have various levels of significance to the practitioners of science at the time. An archetypal example is the paradigm shift from Newtonian classical mechanics to the Einstein's relativity theory.

Engineering science is focused on human needs. Engineering science is inherently creative; it takes what nature provides in order to make something new, or to make things behave in novel ways. Engineering is frequently depicted as being dependent on science and is called "practical science." On the other hand, scientific experiments often rely on engineered experimental instruments. Without engineering, many sciences are impossibly impractical. Engineering is what is essential when pursuing the application of scientific discoveries (McCarthy, 2012). As Sunny Auyang put it: "Natural scientists discover what was not known. Engineers create what did not exist." Science is discovery and engineering is invention.

Social science refers to the studies concerned with society and human behavior. It covers anthropology, archaeology, criminology, economics, education, history, linguistics, communication studies, political science, international relations, sociology, human geography, and psychology, and includes elements of other fields as well, such as law, cultural studies, environmental studies, and social work.

In social science we study personalities as we interact with people of different personalities. Knowledge of personalities will allow us to work efficiently and live pleasantly. A basic task in the investigation into personality is to group people on the basis of similar personalities—an application of the general similarity principle. Such groupings allow us to rationalize a person's extreme behavior since one is viewed as a member of the group rather than an isolated individual. Science is nothing but an application of the similarity principle, while such applications involve proximity and subjectivity. That is why proximity and subjectivity are the essential characteristics of all sciences.

To discuss this further in the context of social science, the norms in a society are naturally formulated based on the opinions and behaviors of the majority in the society. Those who behave extremely outside the norms and in a way harmful to society will be punished. Later, as time passes, people who evidence such bad behavior accumulate and become a group. Until then, scientists (e.g., psychologists) start to study the common cause of their unpleasant actions. When such medical or psychological causes are identified, we rationalize their behavior from, e.g., a medical or neurological perspective. People who have displayed some unpopular personality traits (not necessarily bad) might have been mistreated by society until they became a large group, large enough so that a new science (a science with an ethical component) is developed for such individuals.

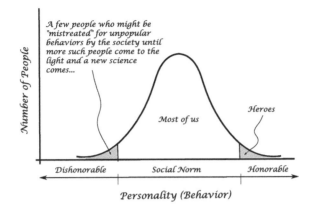

Science has its limits. We cannot know everything, or observe or explain everything about the physical world. The fancy version of this critique points to Heisenberg's uncertainty principle, Fitch's knowability paradox, and Gödel's Incomplete Theorem.

5.2 Evolution

The chicken or the egg paradox we all know is "Which came first, the chicken or the egg?" This question also evokes a more general question of how life and the universe began.

The *Theory of Evolution* answers the question as follows: Species change over time via mutation and selection. Since DNA can be modified only before birth, a mutation must have taken place at conception or within an egg so that an animal similar to a chicken, but not a chicken, laid the first chicken egg. Thus, both the egg and the chicken evolved simultaneously from birds that were not chickens and did not lay chicken eggs but gradually became more and more like chickens over time.

According to Darwin (1859), "... if variations useful to any organic being do occur, assuredly individuals thus characterized will have the best chance of being preserved in the struggle for life; and from the strong principle of inheritance they will tend to produce offspring similarly characterized. This principle of preservation, I have called, for the sake of brevity, Natural Selection."

Darwin implied here the four essential conditions for the occurrence of evolution by natural selection:

(1) *Reproduction* of individuals in the population,

(2) *Heredity* in reproduction,

(3) *Variation* that affects individual survival, and

(4) *Finite resources* causing competition.

For decades, Mendel's ideas were considered to be opposed to Darwin's. *Darwin's theory* asserted that evolution, and therefore variation, is continuous and *Mendel's theory* proposed that variation due to genetic crossover in mating is discrete. The cornerstones of Darwin's theory were that individual variations must be very small, that natural selection on these tiny variations is what drives evolution, and that evolution is gradual. The false opposition between the early Darwinists and Mendelians was cleared up by the 1920s when it was discovered that, unlike the traits of Mendel's peaplants, most traits in organisms are determined by many genes, each with several different alleles. The huge number of possible combinations of these many different alleles can result in seemingly continuous variation in an organism. Discrete variation in the genes of an organism can result in continuous-seeming variation in the organism's phenotype—the physical traits (e.g., height, skin color) resulting from these genes (Mitchell, 2009).

After necessary statistical tools development in the 1920s and 1930s for analyzing the results of Mendelian inheritance with many interacting genes operating under natural selection in a mating population, Darwinism and Mendelism are now recognized as being complementary, not opposed. This unification of the two theories, along with the framework of population genetics, is now called *modern synthesis*, which includes the following principles of evolution: (1) Natural selection is the major mechanism of evolutionary change and adaptation. (2) Evolution is a gradual process, occurring via natural selection with very small individual variation due to random genetic mutations and recombinations (crossover). (3) The observed evolutionary macroscale phenomena, such as the origin of new species, can be explained by the microscopic process of gene variation and natural selection.

The modern synthesis is challenged over the emphasis on gradualism and over the claim that microevolution is sufficient to explain macroevolution. Some evolutionary biologists believe that natural selection is an important mechanism of evolutionary change, but the roles of historical contingency and biological constraints are also undeniably important. They suggest that evolutionary theory be modified to incorporate mechanisms that occur at levels higher than the population (e.g., species sorting). These scientists advocate an extension called hierarchical theory (Mayr, 1980).

Diversity is a necessary condition for evolution. However, more diversity can either speed up evolution or cause chaos in the population. Human society could become homogeneous owing to interacial marriages, the Internet, promotion of social equalities, and other factors. Such homogenization slows down evolution. Nevertheless, humans are still evolving. Experts believe that about 9 percent of our genes are undergoing rapid evolution, nearly as we

Cross-Species Evolution

speak! The genes most affected by natural selection are those involving the immune system, sexual reproduction, and sensory perception.

However, not every scientist believes in evolutionary theory. Some completely oppose it and others agree on intraspecies evolution but contest cross-species evolution.

The notion of natural selection has been used (and abused) throughout many scientific fields and in our daily lives.

Artificial selection, a major technological application of the evolution principles, is the intentional selection of certain traits in a population of organisms. Humans have used artificial selection for thousands of years in the domestication of plants and animals, and more recently in genetic engineering, using selectable markers such as antibiotic resistance genes to manipulate DNA in molecular biology.

Theoretically, evolution can happen on multiple levels: cell evolution makes better cells, organ evolution make stronger organs, and human evolution makes heathy, happy, and longer-lived humans. But how can we be sure these nested evolutions will not be in conflict. We have seen that tumor cells are very strong in competing for nutrition with normal cells. However, such strong tumor cells are definitely miserable from the perspectives of organ and human evolutions. Our society, the consequence of numerous individually motivated actions, is a complex system supported by swarm intelligence. The natural evolution can make each of us more compatible with our neighbor, but this does not necessarily make any individual or society better, as seen by the Braess paradox (Section 6.2) and the shootout game (next section).

5.3 Devolution

The concept of *devolution* or degenerative evolution was proposed by scientists in the 19th century, though beliefs in evolution were in the main stream at that time. One of the first biologists to suggest devolution was Ray Lankester. Lankester explored the possibility that evolution by natural selection may in some cases lead to devolution; an example he studied was the regressions in the life cycle of sea squirts. He was a critic of progressive evolution, pointing out that higher forms existed in the past which have since degenerated into simpler forms. Lankester argued that "if it was possible to evolve, it was also possible to devolve, and that complex organisms could devolve into simpler forms or animals" (Creed, 2009, p. 8; Moore, 2002, p. 117).

In 1857 the physician Bénédict Morel claimed that environmental factors such as taking drugs or using alcohol would produce degeneration in the off-spring of those individuals, and would revert those offspring to a primitive state (Moore, 2002). Using Darwin's theory and many rival biological accounts of development then in circulation, scientists suspected that it was just as possible to devolve, to slip back down the evolutionary scale to prior states of development (Luckhurst, 2005).

There is no reason to believe that devolution never happens within and between the same organisms. In fact, it can occur at different levels for various reasons: (1) A medicine can cure disease and make people weaker longer, while at the same time devolving the human's immune system. (2) As the environment changes, an organism that fits well in one environment may not fit into another; thus, as environment circularly alters (like four seasons), the best fitting organism alters accordingly. (3) Evolution can occur at a higher level such as society and devolution can happen at lower levels such as with individuals and vice versa.

Since "fitness" is multifacted, defining it is often subjective and difficult. For instance, suppose one couple are healthy and are expected to live long, but have low fertility, whereas another couple are not as healthy as the first couple, but have a high fertility and decide to have more children. Which couple will be the winner via evolution after generations? Indeed, people have criticized the Darwinian notion of "survival of the fittest" by declaring that the whole thing is a simple tautology: whatever survives is "fit" by definition! Defenders of the notion argue that fitness can be quantified by empirical measures such as speed, strength, resistance to disease, and aerodynamic stability (for our flying friends) independent of survivability (Gintis, 2009). Nevertheless, under some conditions it may be simply false as shown in the following game, adapted from Shubik and Shapley (1954).

Suppose Andy, Bob, and Charlie are going to have a shootout. In each round, until only one player remains standing, the current shooter can choose one of the other players as a target and is allowed one shot. At the start of the game, straws are drawn to see who goes first, second, and third, and they take turns repeatedly in that order. The goal for each person is to stay alive as long as possible. Each player can shoot a person or shoot in the air. A player who is hit is eliminated. Andy has 50% accuracy, Bob has 80% accuracy, and Charlie is a perfect shooter with 100% accuracy.

The Least Accurate Shooter Will Survive Longest

There are three scenarios: (1) People don't know and will not learn in the game other people's shooting skills; thus, they decide randomly to shoot as accurately as possible at a person. (2) Everyone knows everyone else's shooting skills, he always shoots the most threatening person alive. (3) Bob and Charlie always shoot at each other, but Andy will pretend to be innocent, shooting in the air until one of them has died, then he shoots at the remaining person. Probability theory and computer simulations show that in Scenario 1 (shooting at people randomly), the survivability or chance of surviving for Andy, Bob, and Charlie is 21.7%, 34.3%, and 44%, respectively. However, in Scenarios 2 and 3, the survivability of the players is surprisingly, inverse to their accuracy, 45% for Andy, 31% for Bob, and 24% for Charlie in Scenario 2 and 52.2% for Andy, 30% for Bob, and 18.2% for Charlie in Scenario 3. Both Bob and Charlie have smaller survival chances than random shooting. They would be better off if they could cooperate, that is, eliminate Andy first. In such case, Andy should not take the strategy in Scenario 3.

You may think that if the worst shooter survives longest, why cannot the best shooter sometimes shoot in the air so that he can become the worst shooter and survive longer? But that does not work because the best shooter (the skill of the person not what he actually does) is considered the most threatening one by others. Therefore, everyone who wants to survive longer will have to do his best to eliminate the most threatening person, and then the second most threatening person. As a result, the least threatening (least accurate) shooter will survive longest in this society.

Situations like the shooting game happen anytime everywhere: in political elections, the voters can be divided into 3 (or more) groups: the best "shooter," the good "shooter," and the bad "shooter." Of course, the classification of the subjects into groups is not unique, which make the problem more interesting. Mao Zedong who ruled China for 27 years (1949–1976) and developed the *Three Worlds Theory*, posited that international relations comprise three politico–economic worlds: the First World, the superpowers (developed countries), the Second World, the superpowers' allies (developing countries), and the Third World, the nations of the Nonaligned Movement (underdeveloped countries). He emphasized that China as a Third World country should join the force of the Second World against the First World. Notably, the United States has a different division of the Three Worlds.

The point here is that *micro-motivative behavior* at the individual level can be a devolutionary force, making a weaker person live longer and the society devolve in a sense. Controversially, everyone, every member of the human race, should (ethically) have an equal chance to live for an equally long time. This is a force against evolution, we may argue, that will however make a better society.

5.4   Classical Game Theory

Game Theory is a distinct and interdisciplinary approach to the study of human behavior. *Game Theory* addresses interactions using the metaphor of a game: In these serious interactions, the individual's choice is essentially a choice of strategy, and the outcome of the interaction depends on the strategies chosen by each of the participants. The significance of Game Theory is dignified by the three Nobel Prizes to researchers whose work is largely in Game Theory: in 1994 to Nash, Selten, and Harsanyi, in 2005 to Aumann and Schelling, and in 2007 to Maskin and Myerson.

Tucker's *Prisoners' Dilemma* is one of the most influential examples in economics and the social sciences. It is stated like this: Two criminals, Bill and John, are captured at the scene of their crime. Each has to choose whether or not to confess and implicate the other. If neither man confesses, then both will serve 1 month. If both confess, they will go to prison for 5 months each. However, if one of them confesses and implicates the other, and the other does not confess, the one who has collaborated with the police will go free while the other will go to prison for 10 months.

The strategies offered in this case are confess or don't confess. The payoffs or penalties are the sentences served. We can express all this in a standard Game Theory *payoff table*.

	John Confesses	John Doesn't
Bill Confesses	$(5,5)$	$(0,10)$
Bill Doesn't	$(10,0)$	$(1,1)$

The Prisoner's Dilemma

Let's discuss how to solve this game. Assume both prisoners are rational and try to minimize the time they spend in jail. Bill might reason as follows: If John confesses, I will get 10 months if I don't confess and 5 months if I do, so in that case it's best to confess. On the other hand, if John doesn't confess, I will go free if I confess and get 1 month if I don't confess. Therefore, either way, it's better (best) if I confess. John reasons in the same way. Therefore, they both will confess and serve 5 months in jail. This is a solution or equilibrium for the game. This is a *noncooperative solution*, a solution to a noncooperative game in which no communication is allowed and thus no collaboration exists between the players. However, we can see in this example that it is better for both players if they both don't confess, which is a solution of a cooperative game.

The pair of strategies John and Bill are willing to take constitute the so-called *Nash Equilibrium*, an important concept in Game Theory. Nash Equilibrium, named after John Forbes Nash, is a solution concept of a game involving two or more players in which no player can benefit by changing his strategy unilaterally. If each player has chosen a strategy and no player can benefit by changing his strategy while the other players keep theirs unchanged, then the current set of strategy choices and the corresponding payoffs constitute a Nash Equilibrium (Nash, 1951). The Nash Equilibrium is a pretty simple idea: We have a Nash Equilibrium if each participant chooses the best strategy, given the strategies chosen by other participants. In the Prisoners' Dilemma, the pair of strategies {confess, confess} constitutes the equilibrium of the game.

Most classic game theories assume three conditions: *common knowledge*, *perfect information*, and *rationality*. A fact is common knowledge if all players know it, and know that they all know it, and so on. The structure of the game is often assumed to be common knowledge among the players. A game has perfect information when at any point in time only one player makes a move and knows all the actions that have been made until then. A player is said to be rational if he seeks to play in a manner that maximizes his own payoff. It is often assumed that the rationality of all players is common knowledge. A payoff is a number that reflects the desirability of an outcome to a player, for whatever reason. When the outcome is random, payoffs are usually weighted with their associated probabilities to form the so-called expected payoff that incorporates the player's attitude toward risk.

Games in which the participants cannot make commitments to coordinate their strategies are known as *noncooperative games*. In a noncooperative game, the solution from an individual perspective often leads to inferior outcomes, as in the Prisoners' Dilemma. In a noncooperative game, the rational person's problem is to answer the question "What is the rational choice of a strategy when other players will try to choose their best responses to my strategy?" In contrast, games in which the participants can make commitments to coordinate their strategies are *cooperative games*.

It is preferable, in many cases, to define a criterion to rank outcomes for the group of players as a whole. The Pareto criterion is one of this kind: an outcome is better than another if at least one person is better off and no one is worse off. If an outcome cannot be improved upon, i.e., if no one can be made better off without making somebody else worse off, then we say that the outcome is Pareto Optimal.

In the real world, a *Pareto optimal outcome* for a cooperative game is usually not unique. The set of all Pareto-Optimal outcomes is called the solution set, which is not very useful in practice. To narrow down the range of possible solutions to a particular price or, more generally, distribution of the benefits, is the so-called *bargaining problem*. The range of possible payments might be influenced, and narrowed, by competitive pressures from other potential suppliers and users, perceived fairness, and bargaining.

A group of players who commit themselves to coordinate their strategies is called a *coalition*. The standard definition of efficient allocation in economics is *Pareto optimality*. An allocation is said to be dominated if some of the members of the coalition can do better for themselves by deserting that coalition for some other coalition. *Core* is an important concept in game theory. The core of a cooperative game consists of all allocations with the property that no subgroup within the coalition can do better by deserting the coalition.

5.5   Evolutionary Game Theory

If Bill and John play the Prisoner's Dilemma many times, common sense tells us that the players will cooperate eventually, and this intuition is supported by experimental evidence (Andreoni and Miller 1993). Perfectly "rational" individuals fail to play the Nash equilibrium in this case because it is unintelligent for Bill to defect on every round. Indeed, with a little bit of thought, Bill will cooperate on the first couple of rounds, just to see what John will do. John will be thinking the same thing. In reality, it is unlikely that the exact game will be played many times, but everyone plays many similar games many times in his life. A reputation of cooperation is important to the outcome of future games. That is why people tends to play cooperatively or not play defection, especially within their established network.

Axelrod's studies (1984, 1986) of the Prisoner's Dilemma made a big splash starting in the 1980s, particularly in the social sciences. Axelrod added in the game social censure for defecting when others catch the defector in the act. In Axelrod's multiplayer game, every time a player defects, there is some probability that some other players will witness that defection. In the game, each player can decide whether to punish a defector if the punisher witnesses the defection. In particular, each player's strategies consist of two numbers: a probability of defecting (boldness) and a probability of punishing a defection that the player witnesses (Mitchell, 2009).

Any game is a subgame of the big game of life. But subgames and the big game are often inseparable. People in the same network have relatively stable relationships to some strangers you might run into at an airport. That is, you play the repeated (similar) subgames with them and a reputation of cooperation is important.

Evolution of Games of Life

If the game is infinitely repeated this is not very interesting to us since a player can always signify to the other player that he wants to cooperate until the other player gets it. A player's expected payoff will not change when he plays a finite number of times using different strategies before he takes the final strategy. This is because the expected value is determined by the final strategy that he will employ an infinite number of times.

The common knowledge assumption from classical game theory is often not true. Player one has a perception of the payoff matrix of player two, which might be significantly different from what player two actually has in his mind. People even have different payoff matrices in their minds, each associated with a probability.

The timing of the willingness to cooperate is also important. Imagine that some players play a cooperative game, signify others, but feel hopeless and start to play a noncooperative game before other players (slow newcomers) start to play a cooperative game. Those slow newcomers also feel hopeless because others play no more cooperative games. In the real world, people have different attitudes towards risk, which causes asynchronized cooperation, and effectively leads to a noncooperative game.

Evolutionary game (EG) is the study of repeated games and how such games evolve over time. A key property of EG is selection dynamics.

Let's denote the relative frequency of the i-th type of populations (who play strategy i) of the t-th generation by $x_i(t)$. The *selection dynamics* is the equation that characterizes the population change in i-th type player from t-th generation to $(t + 1)$-th generation:

$$x_i(t + 1) - x_i(t) = x_i(t)\left(f_i - \phi\right),$$

where f_i is called the fitness for the i-th type and the average fitness $\phi = \Sigma_i x_i f_i$ is a constant. If $f_i > \phi$, the population for the i-th type will increase; otherwise, it will remain constant or decrease.

Under the condition of constant population size and constant fitness f_i, the system shows competitive exclusion: the fittest type will outcompete all others—"survival of the fittest." When f_i is a function of time and population characteristics and includes random errors caused by e.g., mutation and strategy shift, it becomes an evolutionary game. In evolutionary games, fitness is often a linear function,

$$f_i(t) = \Sigma_j x_j(t)a_{ij},$$

where a_{ij} is the payoff matrix of the game. A linear fitness is often more reasonable than a constant fitness since a larger group size will be helpful in the competition for survival. The selection dynamics with a linear fitness is a nonlinear equation, wherein all manner of strange phenomena can happen over time, including chaotic behavior.

Now the question is: are we acting rationally? My answer is: We all live in a large game with many subgames, where most of us act irrationally (being somewhat nearsighted, just like ants), only a few act rationally at times. Yes, there are a majority of ants and only a few who play classic games.

5.6   Networks and Graph Theory

A *network* is a collection of members (called nodes) that have relatively stable interrelationships (called links). Such stable interrelationships allow regular "information" exchanges among its members. Such information exchange can lead to the changes of the member's behavior and eventually affect the topology and dynamics of the network. For example, each of use has friends who, in sum, form our friend network. Such a relatively stable group allows us to share "information," more often within the network than outside the network. The information exchanged can take many forms; the passing on of disease is but one unusual example. There are different networks that can be formed and used for different purposes, for instance, the Internet, social, transportation, biological, communication, electronic, and electrical networks, just to list a few. A network can be artificial, such as an artificial neural network that mimics human neural networks and can be used to effectively solve practical problems.

A *social network* is one of the most interesting networks since individual nodes benefit (or suffer) from direct and indirect relationships. Friends might provide access to favors from their friends. Just as Jackson (2008, p. 3) describes: "Social networks permeate our social and economic lives. They play a central role in the transmission of information about job opportunities and are critical to the trade of many goods and services. They are the basis for the provision of mutual insurance in developing countries. Social networks are also important in determining how diseases spread, which products we buy, which languages we speak, how we vote, as well as whether we become criminals, how much education we obtain, and our likelihood of succeeding professionally. The countless ways in which network structures affect our well-being make it critical to understand (1) how social network structures affect behavior and (2) which network structures are likely to emerge in a society."

Biological networks (pathways) are interesting to many scientists. A *biological pathway* is a molecular interaction network in biological processes. The pathways can be classified into fundamental categories: regulatory, metabolic, and signal transduction pathways. There are about 10,000 pathways in humans, nearly 160 pathways involving 800 reactions.

A *gene regulatory pathway* or *genetic regulatory pathway* is a collection of DNA segments in a cell which interact with each other and with other substances in the cell, thereby governing the rates at which genes in the network are transcribed into mRNA. In general, each mRNA molecule goes on to make a specific protein with particular structural properties. The protein can be an enzyme for breakdown of a food source or toxin. By binding to the promoter region at the start of other genes, some proteins can turn the genes on, initiating or inhibiting the production of another protein.

A *metabolic pathway* is a series of chemical reactions occurring within a cell, catalyzed by enzymes, resulting in either the formulation of a metabolic product to be used or stored by the cell or the inhibition of another metabolic pathway. Pathways are important to the maintenance of homeostasis within an organism.

A *signal transduction pathway* is a series of processes involving a group of molecules in a cell that work together to control one or more cell functions, such as cell division or cell death. Molecular signals are transmitted between cells by the secretion of hormones and other chemical factors, which are then picked up by different cells. After the first molecule in a pathway receives a signal, it activates another molecule, and then another until the last molecule in the signal chain is activated. Abnormal activation of signaling pathways can lead to a disease such as cancer. Drugs are being developed to block these disease pathways.

Systems biology is a newly emerging, multidisciplinary field that studies the mechanisms underlying complex biological processes by treating these processes as integrated systems of many interacting components (Materi and Wishart, 2007).

Networks often share certain common structures. Network topology is the study of static properties or structures of a network. What follows is some basic terminology for the topology of network. A walk is a sequence of links connecting a sequence of nodes. A cycle is a walk that starts and ends at the same node, with all nodes appearing once except the starting node, which also appears as the ending node. A path is a walk in which any given node appears at most once in the sequence. A tree is a connected network that has no cycle. In a tree, there is unique path between any two nodes. A forest is a network with tree components. A component is any connected network.

How many friends (*links*) a person has often reflects the importance of the person (called *centrality*) in the network, which can be characterized by the *degree of a node*—the number of direct links a node has. *Authority* is the number of nodes that point to a given node. *Hub* is the number of nodes to which a node points. We often need to know how fast information can travel from one person to another, which is described by *geodesics*—the shortest path between the two nodes. The centrality of a node can also be measured by its

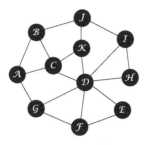

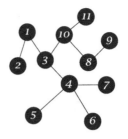

C-D-E-F-G-D-H is a walk, but not a path
because it includes a cycle, D-E-F-D.

A Tree: A graph component
with no cycle

Graphical Components

so-called *closeness*, which is defined as the inverse of average geodesics from this node to all the others in the network. *Closeness* measures how easily a node can reach other nodes, whereas the *diameter of a network*, i.e., the largest distance between any two nodes in the network, can be an indicator of the influential range of a person (node).

One of the mathematically simplest network models is the so-called *random graph*, in which each node is connected (or not) with an independent probability of having a binomial distribution of degrees. However, most networks in the real world, e.g., the Internet, the World Wide Web, biological networks, and some social networks, are highly right-skewed, meaning that a large majority of nodes have low degree, but a small number, known as "hubs," have high degree. This means that there is only a relatively small number of nodes playing the big roles. Similarly, the distribution of geodesics is a measure of the closeness of a network. Calculating the geodesics' distribution is straight forward. Denote a direct connection between nodes i and j by g_{ij}; $g_{ij} = 1$ if i and j are connected; otherwise $g_{ij} = 0$. Then matrix $[g^k]$ represents all possible paths with length k in the network. For example, $[g^2]_{ij} = \sum_k g_{ik}g_{kj} = 3$ implies there are 3 walks of length 2 between nodes i and j, while $[g^3]_{ij} = \sum_k \sum_m g_{ik}g_{km}g_{mj} = 4$ means there are 4 walks of length 3 between nodes i and j.

Some relationships in a network have direction and other properties that are necessary to model real world problems. A *Petri net* (Petri, 1962) is a mathematical network model for simulating a dynamic system such as electronic signal processes, traffic systems, and biological pathways (Chang, 2010).

5.7 Evolutionary Dynamics of Networks

A network can change over time. In a random graph, each node has an equal probability of getting connected to other nodes regardless of its degree, the number of connections it has already. In another kind of network, called *Richer gets Richer*, a node with a higher degree is more likely to get new connections. In yet another type of network a node with a lower degree more often gets new connections.

As elucidated earlier a network has a set of relationships that are relatively stable. We may study a disease spread through a random connection, such as by meeting people on elevators and people passing by each other on the street or attending a conference, where the relationships are temporary. The method to deal with such random interactions is the classic diffusion equation (for an infinite population) or the logistic equation (for finite populations). Regarding the members (nodes) in a network, their interactions are not completely random due to the network structure. The dynamics of a network is network-topological dependent. For instance, in biological networks, Petri net dynamics include *reachability, liveness,* and *boundedness.* Reachability study concerns whether a particular (marking) state can be reached from the initial state; liveness study is to address if any transition will eventually be fired for the given initial state; whereas boundedness study is to investigate if the number of (energy) tokens at each place (node) is bounded for any initial marking of the Petri net (Chang, 2010).

Evolutionary graph theory is an area of research lying at the intersection of graph theory, probability theory, and mathematical biology. Evolutionary graph theory is an approach in studying how the underlying topology of a network affects the evolution of a population (Lieberman, Hauert, and Nowak, 2005; Nowak, 2006).

Evolutionary dynamics act on populations in networks. In small populations, random drift dominates, whereas large populations are sensitive to subtle differences in selective values. Classical evolutionary dynamics was studied in the context of homogeneous or spatially extended populations. In studying evolutionary dynamics of networks, individuals are arranged at the vertices of the network. A homogenous population corresponds to a fully connected graph and spatial structures are represented by lattices where each node is connected to its nearest neighbors.

An earlier model for the dynamics of selection and random drift is the so-called *Moran process*:

(1) Consider a homogeneous population of size N consisting of residents and *mutants.*
(2) At each time step an individual is chosen for reproduction with a probability proportional to its fitness (here, only residents are selected for reproduction).
(3) A randomly chosen individual is eliminated (killed off).
(4) The offspring replaces the eliminated individual.

The Moran process has two possible equilibrium states, called *fixations*: the population consists of either all residents or all mutants. The Moran process describes the *stochastic evolution* in a finite population of constant size. Suppose all N resident individuals B are identical in a completely connected network and one new mutant A is introduced. The new mutant has relative fitness r, as compared to the residents, whose fitness is 1. The *fixation probability*,

$$R_1 = \frac{1 - 1/r}{1 - 1/r^N},$$

is the probability that a single, randomly placed mutant of type A will replace the entire population B.

According to the *isothermal theorem*, a graph has the same fixation probability as the corresponding Moran process if and only if it is isothermal (e.g., a fully connected graph), so that the sum of all weights that lead into a vertex is the same for all vertices.

In general, individuals occupy vertices of a weighted directed graph and the weight w_{ij} of an edge from vertex i to vertex j denotes the probability of i replacing j. The weight corresponds to the biological notion of fitness where fitter types propagate more readily. The temperature of a node is defined as the sum over the weights of all incoming links and indicates how often it is replaced, i.e. a hot node is replaced often and a cold node is rarely replaced.

Networks (graphs) can be classified into *amplifiers of selection* and *suppressors of selection*. If the fixation probability of a single advantageous mutation is higher than the fixation probability of the corresponding Moran process, then the graph is an amplifier, otherwise a suppressor of selection. An example of *evolutionary suppressors* is a central hub that is connected to the nodes along the periphery but the peripheral nodes have no outgoing links (the left figure below) because only the mutant (the central node) can replace the residents—the population becomes all mutants with certainty. Generally, any graph with a single root or small upstream population that feeds into large downstream populations acts as a suppressor of selection. On the other hand, it is also possible to create population structures that amplify selection and suppress random drift. An example of evolutionary amplifiers would be on the star structure, where all nodes are connected to a central hub and vice versa (the right figure), the fixation probability of a randomly placed mutant becomes (Nowak, 2006)

$$R_2 = \frac{1 - 1/r}{1 - 1/r^{2N}}$$

for a large number of residents, N. Thus, any relative fitness r is amplified to r^2.

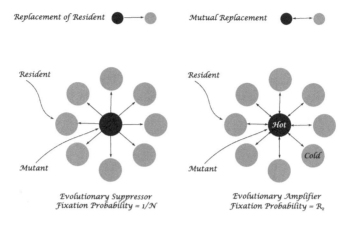

Fixation of Network Dynamics

The study of game theory in the network setting is called network evolutionary game theory. The game players interact with each other through network connections. In such interactions, your friends' strategies (thus their gain or loss) will affect your strategy. Thus, a "good guy" can become a "bad guy" or vice versa, with various probabilities. The dynamics of network games can be studied through Monte Carlo (computer) simulations.

5.8   Brownian Motion

Brownian motion was one of the early crucial discoveries concerning the discreteness of matter. First observed by Jan Ingenhousz in 1785, and later rediscovered by Brown in 1828, the phenomenon consists of the erratic or fluctuating motion of a small particle when it is embedded in a fluid (Parrondo and Dinis, 2004). The quantitative explanation was given by Einstein 1905. Brownian motion is also called the Wiener process in recognition of Norbert Wiener's mathematical study of the phenomenon. Brownian motion is commonly seen in daily life by those who are curious enough to look. For instance, the movement of dusty particles in the air is Brownian motion in three-dimensional space. A random walk can be considered as a two-dimensional Brownian motion.

The modern theory of Brownian motion provided a way to reconcile the paradox between two of the greatest contributions to physics in the second half of the nineteenth century: thermodynamics and the kinetic theory of gases. Key to kinetic theory was the idea that the motion of individual particles obeyed perfectly reversible Newtonian mechanics. In other words there was no preferred direction of time. But the second law of thermodynamics expressly demanded that many processes be irreversible (Haw, 2005). Einstein's theory of Brownian motion explains why many processes are irreversible, in a probabilistic sense.

Brownian Motion

Brownian motion has some fantastic paradoxical properties, and here are three:

(1) Mathematically, Brownian motion is a motion with a continuous path that is nowhere differentiable with probability 1. This means that the instantaneous speed cannot be defined anywhere. This is because, according to Einstein's theory, the mean distance traveled by the particles is proportional to the square root of the time t. By taking the derivative of the mean distance with respect to t, the obtained mean speed is proportional to $1/\sqrt{t}$. From this it follows that the instantaneous ($t = 0$) speed of the particles would be infinite at any moment (Székely, 1986).

(2) The trajectories (realizations) of Brownian motion are rather irregular since they are nowhere differentiable (smooth). In the usual sense, we consider any irregular curve, such as the trajectory of planar Brownian motion, one dimensional. At the same time, it can be shown that the trajectory of a planar Brownian motion actually fills the whole plane (each point of the plane is approached with probability 1). Therefore, the trajectories can also be considered as two-dimensional curves.

(3) Brownian motion has the feature of self-similarity, i.e., the average features of the function do not change while zooming in, and we note that the trace of a Brownian function zooms in quadratically faster horizontally (in time) than vertically (in displacement). More precisely, if $B(t)$ is a Brownian motion, then for every scaling $c > 0$ the process $\tilde{B}(t) = B(ct)/\sqrt{c}$ (the scaling law for fractal objects) is another Brownian process. The self-similarity property is also called scale invariance and is the fundamental characteristic of fractals.

Brownian motion is *memoryless motion*; i.e., the future is only dependent on the present not the past.

Scaling laws for fractal objects state that if one measures the value of a geometric characteristic $\theta(w)$ on the entire object at resolution w, the corresponding value measured on a piece of the object at finer resolution $\theta(rw)$ with $r < 1$ will be proportional to $\theta(w)$:

$$\theta(rw) = k\theta(\omega)$$

where k is a proportionality constant that may depend on r.

The above-delineated dependence of the values of the measurements on the resolution applied suggests that there is no single true value of a measured characteristic. A function $\theta(\omega)$ satisfying the scaling law is the power law function: $\theta(\omega) = \beta\omega^{\alpha}$.

Fractals are closely related to scale-free networks by the power law. Many people believe that World Wide Web links, biological networks, and social networks are scale-free networks. The scale-free property strongly correlates with the network's robustness to failure.

In Brownian motion, the random movements of small particles at microscopic levels result in *diffusions* macroscopically. We see diffusion everywhere, heat transport from a place of higher temperature to a place of lower temperature, drug distribution in blood, and even the spread of technology from the first world to the third world. Globalization in its literal sense is the transformation of local or regional phenomena into global ones. The transformation takes place simply because there are differences among regions. Globalization is a process by which the people of the world are unified into a single society and function together. This process is a combination of economic, technological, sociocultural and, political forces. Despite the complexity of the underlying reasons at the microscope (e.g., collisions among small particles or interactions among different people), the overall movement of globalization can be modeled using a *diffusion equation* at the macroscopic level.

5.9   Stochastic Decision Process

Among the simplest and most widely studied and applied *stochastic processes* is the *Markov chain*, named after the mathematician Andrey Markov. In a Markov chain the probability of transitioning from one state of a process to another does not depend on how we got to the current state. A *Markov decision process* (MDP) is similar to a Markov chain, but with MDPs there are also actions and rewards. MDPs provide a powerful mathematical framework for modeling decision-making processes in situations where outcomes are partly random and partly under the control by the decisionmaker. MDPs are useful in studying a wide range of optimization problems solved via dynamic programming and reinforcement learning (e.g., the training of robots). Since the 1950s when the concept was first explored (Bellman 1957; Howard, 1960), MDPs have been widely used in robotics, automated control, economics, business management, nursing system, manufacturing, and pharmaceutical industry research (Chang, 2010).

When a system reaches a steady state, the *transition probability* is only dependent on the two states involved and the action taken, a, but is independent of the time when the transition occurs.

Let's denote a dynamic system which moves over states $s_1, s_2, ..., s_N$ and the motion is controlled by choosing a strategy or a sequence of actions $a_1, a_2, ..., a_N$. There is a numerical reward $g_i(s, a_i)$ associated with each action a_i ($i = 1, 2, ..., N$) at state s. The goal is to find a policy (strategy) which maximizes the total expected gain.

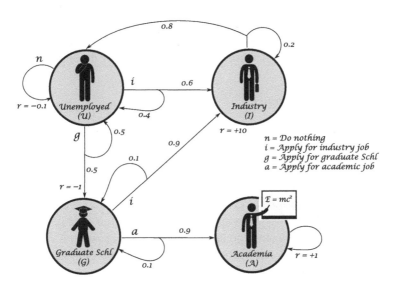

What Is the Best Policy in the Career Options

An example of a Markov decision process from an anonymous university lecture handout about undergraduates' career options is presented in the figure above, where r is the reward and an agent initial state is U unemployed. The rewards and transitional probabilities, or even the model itself, may vary from individual to individual, but the model has well illustrated the fact that many decision problems we face in our life's journey can be modeled by MDPs. The assumed transition probabilities between the four states $\{U, I, G, A\}$ are marked in the figure. For instance, if a person takes action g (applies for graduate school) while in the state U, his immediate reward is $r = -1$; if he does nothing (action n), his reward is $r = -0.1$. If he takes action i (applies for an industry job), he will have a 0.6 probability of getting a job and a reward of $r = +10$.

An effective way to solve an MDP problem is backward induction. When Bellman's principle of optimality is applied to a stochastic decision problem, the optimal policy is regarding the expected gain (or loss) as opposed to a loss in a deterministic case.

Computationally, the stochastic decision problems are usually solved using the so-called *Bellman Equations*:

$$V(s) = r(s, a) + \Pr(w|s, a) V(w),$$

where $V(s)$ is the expected reward at state s, $r(s, a)$ is the instant reward by taking action a at state s, and $\Pr(w|s, a)$ is the transition probability from state s to state w when action a is implemented. The *optimal policy* $\pi^* = \{a_1^*, a_2^*, ..., a_N^*\}$ is the sequence of actions that maximize the expected reward $V(s)$ at the current state s (see Chang 2010 for details).

5.10   Swarm Intelligence

According to Mitchell (2009), a *complex system* is a system of large networks of components (with no central control) in which each individual in the system has no concept of collaboration with her peers, and simple rules of operation give rise to complex collective behavior, sophisticated information processing, and adaptation via learning or evolution. Systems in which organized behavior arises without a centralized controller or leader are often called *self-organized systems*, while the intelligence possessed by a complex system is called *swarm intelligence (SI)* or *collective intelligence*. SI is an emerging field of biologically inspired artificial intelligence, characterized by micro motives and macro behavior.

The structure of a complex system can be characterized using a network. However, the dynamics are not simply the sum of the static individual components. Instead, the system is adaptive through interactions between the individual components and the external environment. Such adaptations at the macroscopic level occur even when the simple rules governing the behaviors of each individuals have not changed at all.

A good example of SI is that of ant colonies which optimally and adaptively forage for food. Ants are able to determine the shortest path leading to a food source. The process unfolds as follows. Several ants leave their nest to forage for food, randomly following different paths. Ants keep releasing *pheromones* (a chemical produced by an organism that signals its presence to other members of the same species) during the food search process. Such

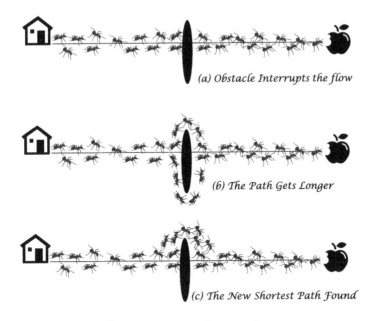

(a) Obstacle Interrupts the flow

(b) The Path Gets Longer

(c) The New Shortest Path Found

Swarm Intelligence: Ants Adapt to the Environment

pheromones on the path will gradually disperse over time. Those ants reaching a food source along the shortest path are sooner to reinforce that path with pheromones, because they are sooner to come back to the nest with food; those that subsequently go out foraging find a higher concentration of pheromones on the shortest path, and therefore have a greater tendency (higher probability) to follow it. In this way, ants collectively build up and communicate information about locations, and this information adapts to changes in the environmental conditions! The SI emerges from the simple rule: follow the smell of pheromones.

Swarm intelligence is not an "accident" but rather a property of complex systems. It does not arise from a rationale choice, nor from an engineered analysis. Individuals in the system have no global perspective or objective. They are not aware of what is globally happening. They are not aware how their behavior will affect the overall consequence. The behavior of swarm intelligent systems is often said to be an *"emergent behavior."*

There are a lot of examples of emergent behaviors: Bee colony behavior, where the collective harvesting of nectar is optimized through the waggle dance of individual worker bees; flocking of birds, which cannot be described by the behavior of individual birds; market crashes, which cannot be explained by "summing up" the behavior of individual investors; and human intelligence that cannot be explained by the behaviors of our brain cells. Likewise, a traffic jam is not just a collection of cars, but a self-organized object which emerges from and exists at a level of analysis higher than that of the cars themselves (Xiaohui Cui, Swarm Intelligence Presentation, U.S. Department of Energy).

In a complex system, an individual agent neither has enough intelligence to solve the problem at hand nor has any goal or intention to solve it. A collection of intelligent agents can produce a better solution to a problem facing the group than the sum of the abilities of all agents when they work individually.

Interestingly, not long ago Southwest Airlines was wrestling with a difficult question: Should it abandon its long-standing policy of open seating on planes (Miller, 2010)? Using computer simulations based on artificial ant colonies, Southwest figured out that the best strategy is to assign the seats at check-in, but boarding would still be first-come, first-served. Now this strategy has become a standard for various airlines. Further applications of SI have been developed in cargo systems and in many other fields.

An SI algorithm is a kind of computer program comprising a population of individuals that interact with one another according to simple rules in order to solve problems which may be very complex. Individuals in an SI system have mathematical intelligence (logical thought) and social intelligence (a common social mind). Social interaction thus provides a powerful problem-solving algorithm in SI.

The swarm intelligence characteristics of a human network integrate two correlated perspectives on human behavior: cognitive space and social space. In SI, we see the evolution of collective ideas, not the evaluation of people who hold ideas. Evolutionary processes have costs: redundancy and futile exploration are but two. But, such processes are adaptive and creative. The system parameters of SI determine the balance of exploration and exploitation. Thus, uniformity in an organization is not a good sign.

An ant is simple, while a colony of ants is complex; neurons are simple, but brains are complex as a swarm. Competition and collaboration among cells lead to human intelligence; competition and collaboration among humans form a social intelligence, or what we might call the global brain. Nevertheless, such intelligence is based on a human viewpoint, and thus it lies within the limits of human intelligence. Views of such intelligence held by other creatures with a different level of intelligence could be completely different!

Swarm intelligence is different from the reinforcement learning. In reinforcement learning, an individual can improve his level of intelligence over time since in the learning process, adaptations occur. In contrast, swarm intelligence is a collective intelligence from all individuals. It is a global or macro behavior of a system. In complex systems there are huge numbers of individual components, each with relatively simple rules of behavior that never change. But in reinforcement learning, there is not necessarily a large number of individuals, in fact there can just be one individual with built-in complex algorithms or adaptation rules.

5.11 From Ancient Pictographs to Modern Graphics

The common purpose of *pictographs*, highly abstract mathematical symbols, and complex data visualization is that they all serve as effective means of thought and communication. The importance of *nomenclature, notation, language,* and *graphics* as tools for thinking cannot be ignored. For example, as Iverson (2013) has pointed out, in chemistry and botany the establishment of systems of nomenclature by Lavoisier and Linnaeus did much to stimulate and channel later investigation.

Language is an instrument of human reason, and not merely a medium for the expression of thought (Boole, 1951). From philology (writing, linguistics), we learn that the use of Chinese characters is one of the world's oldest forms of writing. Starting as early as several thousand years ago, examples can be seen from oracle bone inscription etched in tortoise shell in the Shang Dynasty and clerical writings in the Han Dynasty to regular script in daily life and business today. The evolutionary process of Chinese characters over thousands of years shows the general trend of moving from pictograph towards simplicity (but abstract), not only in simplified character patterns but also in a reduced number of characters.

Mathematical notation provides perhaps the best-known and best-developed example of language used consciously as a tool of thought. Good notation relieves the brain of all unnecessary work, sets it free to concentrate on more advanced problems, and in effect increases the mental power of the race (Cajori, 1929, Iverson, 2013, p.332). We can say that without concise mathematical symbols many profound mathematical discoveries would have been unthinkably difficult. Likewise, without graphics, we would not be able to effectively communicate our keenest observations and ideas, especially those culled from massive data. Thus would many great discoveries in modern science be otherwise impossible, since they require strong collaboration among scientists using graphics.

Scientific graphics is the well-designed presentation of interesting data, a matter of substance, of statistics, and of design (Tufte, 1983). A Chinese proverb says: "A picture is worth a thousand words." Graphics communicate complex ideas in a concise way, with clarity, precision, and efficiency. A graph is that which gives to the viewer the greatest number of ideas in the shortest time with the least ink in the smallest space, is nearly always multivariate, and requires telling the truth about the data (Tufte).

	oracle bone jiaguwen	greater seal dazhuan	lesser seal xiaozhuan	clerkly script lishu	standard script kaishu	running script xingshu	cursive script caoshu	modern simplified jiantizi
rén (*nin) human								人
nǚ (*nraʔ) woman								女
ěr (*nhaʔ) ear								耳
mǎ (*mrāʔ) horse								马
yú (*ŋha) fish								鱼
shān (*srān) mountain								山
rì (*nit) sun								日
yuè (*ŋot) moon								月
yǔ (*whaʔ) rain								雨
yún (*wan) cloud								云

Evolution of Chinese Characters

It is difficult to identify who pioneered the use of graphics in scientific research. The benefits of scientific graphics certainly flow from the contributions of many people in all areas, starting before writing. Their history and evolution can be a great scientific research topic. There are today a growing number of different types of scientific graphics, many of them reflecting humans' remarkable intelligence and creativity. Here we briefly discuss a few that are simple and the most commonly used.

A *scatter plot* is probably the simplest mathematical diagram, using Cartesian coordinates to display values for two variables (x, y) for a set of data. It gives the overview of the distribution of the data. An example is presented in Section 4.1.

A *histogram* is constructed based on tabular frequencies or percentages which are shown as adjacent equal-width rectangles (bins). The heights of the bins are proportional to the frequencies or percentages of the categories in the horizontal (or vertical) axis. A histogram is used to compare frequencies or percentages across several categories. An example is given in Section 4.2.

A *line chart* is a two-dimensional scatter plot of continuous ordered observations where the observations are connected, following their order as a series of data points (x, y), using straight-line or curved segments. A line chart is often used to visualize a trend in data over intervals of time. Examples are provided in Sections 1.5, 3.3, 4.4, 4.5, 6.6, 6.8 and 7.18.

A *tree diagram* is a representation of a tree structure, a way of representing the hierarchical nature of a structure in a graphical form. See the example of a Decision Tree in Section 4.8. More examples can be found in Sections 1.1, 1.9, 3.2, and 3.5.

A *flowchart* is a type of diagram that represents an algorithm or process, showing the steps as boxes of various kinds and their order as indicated by arrows. Flowcharts are used in various fields when analyzing, designing, documenting, or managing a process. Applications of flowcharting can be found in Sections 3.4, 3.9, 4.7, 5.9, 7.8, 7.15, and 7.17.

A *network diagram* consists of nodes and links, where the nodes represent members and the links represents relationships between pairs of nodes. In terms of their links, a network can be directed or undirected or a mixture of both. Examples in this book are presented in, Sections 5.6, 5.7, 6.2, 7.5, 7.6, and 7.7.

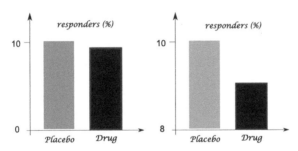

Visual Illusion Due to Different Scale

Graphics give strong visual impact; thus, we should use caution when we create or read them. For instance, the above two graphics presenting the same set of data may give an inexperienced reader very different impressions if he does not pay attention to the different starting points of the vertical axes. A natural question is: Which figure is more appropriate? It is dependent on the particular context. Among many other things, one should also avoid the common mistake of overloading information in a single graph.

Chapter 6

Controversies and Challenges

6.1 Fairness of Social System

A source of great debate in societies throughout the world, both *democratic* and *undemocratic*, is the notion of *fairness*. Most would agree that fairness does not mean we are all guaranteed the right to have or do the same thing in the same circumstance as everyone else. Rather, it often means that everyone is offered the same opportunity to pursue, to do, to gain, to win, etc. But fairness can often be just an illusion. We begin with an example.

Imagine the following game between two opponents A and B. Each player constructs a stack of one-dollar coins. To play the game, each takes out a coin from the top of his stack at the same time. The players cannot flip the coin over. The winning rules are defined as follows:

(1) If heads–heads appears, A wins \$10 from B.
(2) If tails–tails appears, A wins \$0 from B (no winner).
(3) If heads–tails or tails–heads appears, B wins \$5 from A.

In the "fair game," Player A will tend to pile the coins up in such a way as to exhibit heads more often, expecting to find many heads–heads situations. On the other hand, B, knowing that A will try to show heads, will try to show tails. One might think that, since the average gain in the situation that A wins, i.e., $10/2$, and in the situation that B wins, i.e., $5/2 + 5/2$, are the same, it will not really matter how the coins are piled up as long as one opponent does not know the other's stack. However, this is not true. If player B stacks his coins in such a way that one-fourth of them, randomly distributed in the stack, show up as heads, he will expect to win, during a long sequence of rounds, an average of \$1.25 per round. Here is why: Let a (or b) represent the fraction of the stack in which A (or B) puts heads facing up. Since the probabilities of heads–heads, heads–tails, and tails–heads are ab, $a(1-b)$, and

$(1 - a)b$, respectively, the expected gain per round of player B is

$$G = -10ab + 5a\,(1 - b) + 5(1 - a)b.$$

Interestingly, if $b = 0.25$, then $G = \$1.25$ regardless of the value of a.

Game Rules:
(1) If heads–heads appears, Andy wins $10 from Bob.
(2) If tails–tails appears, no winner.
(2) If heads–tails or tails–heads appears, Bob wins $5 from Andy.

Unfairness of a "Fair Game"

In social choice theory, Arrow's *impossibility theorem* (*Arrow's Paradox*) is named after economist Kenneth Arrow (Easley and Kleinberg, 2010, p. 657), who demonstrated the theorem in the original paper "A Difficulty in the Concept of Social Welfare." Arrow was a corecipient of the 1972 Nobel Memorial Prize in Economics. Arrow's theorem is a mathematical result, but it is often expressed in a nonmathematical way by a statement such as "No voting method is fair," "Every ranked voting method is flawed," or "The only voting method that isn't flawed is a dictatorship." These simplifications of Arrow's result should be understood in context. What Arrow has proved is that any social choice system exhibiting three criteria, referred to as universality, unanimity, and independence of irrelevant alternatives (IIA), is a *dictatorship*, i.e., a system in which a single voter possesses the power to always determine the group's preference. In other words, Arrow proved that no voting system can be designed that satisfies these three fairness criteria:

(1) *Universality*: For any set of individual voter preferences, the social welfare function should yield a unique and complete ranking of societal choices.
(2) *Unanimity* (*Pareto efficiency*): If every individual prefers a certain option to another, then this must also be the resulting societal preference.

(3) *Independence of irrelevant alternatives* (IIA): The social preference between A and B should depend only on the individual preferences between A and B (pairwise independence). In other words, a change in individuals' rankings of irrelevant alternatives (e.g., between A and C) should have no impact on the relative societal ranking between A and B.

Various theorists have suggested weakening the IIA criterion as a way out of Arrow's paradox. Proponents of ranked voting methods contend that the IIA is an unreasonably strong condition. Advocates of this position point out that failure of IIA as given above is trivially implied by the possibility of cyclic preferences as illustrated in the following example of ballots:

(1) One third of the vote: A is better than B, which is in turn better than C.
(2) Another third vote: B is better than C, which is in turn better than A.
(3) The final one-third vote: C is better than A , which is in turn better than B.

If these votes are cast, then the pairwise majority preference of the group is that A wins over B, B wins over C, and C wins over A these are rock-paper-scissors preferences for any pairwise comparison. In this circumstance, any aggregation rule that satisfies the very basic majoritarian requirement that a candidate who receives a majority of votes must win the election will fail criterion IIA, if societal preference is required to be transitive (or acyclic).

Efron's Dice

Nontransitive preferences can exist in other situations. For instance, Efron's dice are the four dice A, B, C, D with the following numbers on their six faces: A displaying $\{4, 4, 4, 4, 0, 0\}$, B with $\{3, 3, 3, 3, 3, 3\}$, C having $\{6, 6, 2, 2, 2, 2\}$, and D, $\{5, 5, 5, 1, 1, 1\}$. It can be easily proved (e.g., Chang, 2012) that die A beats die B; B beats C; C beats D, and D beats A, all with the same probability of $2/3$. You may want to try the dice.

So, what Arrow's theorem really shows is that voting is a nontrivial game, and that game theory should be used to predict the outcome of most voting mechanisms. This could be seen as a discouraging result, because a game need not have efficient equilibria; e.g., a ballot could result in an alternative nobody really wanted in the first place, yet everybody voted for.

6.2   Centralized and Decentralized Decisions

From an evolutionary point of view, a society with centralized decision mechanisms (local and national governments) should be better than a social system built in terms of swarm intelligence or micro-motivations. We vote our preferences to the different levels of the governing system, and we hope that our intelligent representatives can make wise decisions for us. In other words, following a good leader seems like the smart thing to do. However, this is not always true. Consider the following paradox (Chang, 2012; Székely, 1986). Suppose that A, B, C, D, and E are the five members of a trial jury. Guilt or innocence for the defendant is determined by simple majority rule. There is a 5% chance that A gives the wrong verdict; for B, C, and D it is 10%, and E is mistaken with a probability of 20%. When the five jurors vote independently, the probability of bringing the wrong verdict is about 1%. Paradoxically, this probability increases to 1.5% if E (who is most probably mistaken) abandons his own judgment and always votes the same as A (who is most rarely mistaken). Even more surprisingly, if the four jurors B, C, D, and E all follow A's vote, then the probability of delivering the wrong verdict is 5%, five times more than that when they vote independently.

From this and the example of swarm intelligence described in Chapter 5, you may wonder if individually motivated decisions may be better than centralized decisions made by our government. However, this is not always true either, as illustrated in the *Braess paradox*.

This paradox, named after the German mathematician Dietrich Braess (Braess, Nagurney, and Wakolbinger, 2005; Braess and Überein, 1969), shows that eliminating a road, rather than building a road, will sometimes improve traffic conditions. When 42nd Street in New York City was temporarily closed to traffic, rather than the expected gridlock, traffic flowed more easily. In fact, as was reported in the September 2, 2002, edition of *The New Yorker*, in the 23 American cities that added the most new roads per person during the 1990s, traffic congestion rose by more than 70% (Havil, 2008).

Suppose that 2000 cars want to travel from A to B (the left figure below) over a weekend. $A \rightarrow C$ and $D \rightarrow B$ are two narrow roads, the travel time for each road increasing with the number of cars on the road. The travel time for x cars travelling on them is $20 + 0.01x$ minutes. $A \rightarrow D$ and $C \rightarrow B$ are two wide roads, the travel time here being 45 minutes regardless of the traffic. $C \rightarrow D$ is a very short wide road with negligible travel time. Because of this, C and D can be considered as in one place. If a traveler takes road $A \rightarrow C$, it will take him 40 minutes at the most, when every traveler takes path $A \rightarrow C$, which still is less than the 45 minutes when travelling on path $A \rightarrow D$. Therefore, everyone will take $A \rightarrow C$. Similarly, at C, everyone will spend 40 minutes on path $C \rightarrow D \rightarrow B$ and no one will take path $C \rightarrow B$. Therefore, the total travel time from A to B will be 80 minutes for every vehicle—a micro-motivated result.

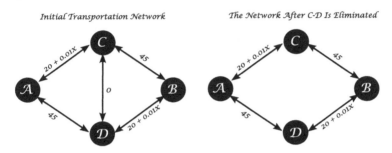

Initial Transportation Network   *The Network After C-D Is Eliminated*

Braess' Paradox in a Transportation Network

Now suppose the road $C \rightarrow D$ is removed by government intervention (the right figure), leaving only two possible paths from A to B: $A \rightarrow C \rightarrow B$ and $A \rightarrow D \rightarrow B$. At the Nash equilibrium (Section 5.4), 1000 travelers will take each of the paths. The travel time for each car is $45 + 20 + 0.01(1000) = 75$ minutes. This result is so surprising because the removal of the optional road $C \rightarrow D$ actually reduces everyone's travel time by 5 minutes or 6.25%. Even more surprising is that a budget cut (in maintaining the road $C \rightarrow D$) in government can actually do something good for everyone!.

The Braess Paradox can be found in most networks, such as water flows, computer data transfer, and telephone exchanges. In 1990, the British Telecom network suffered in such a way when its "intelligent" exchanges reacted to blocked routes by rerouting calls along "better" paths. This in turn caused later calls to be rerouted with a cascade effect, leading to a catastrophic change in the network's behavior (Havil, 2008).

Now a question for you to ponder: Since the noncooperative solution of 80 minutes in travel time is the result when every driver knows the travel condition of every path, do you think any device that provides comprehensive information about travel conditions will help ease the overall travel condition or make it worse?

6.3 Newcomb's Paradox

Newcomb's Paradox came about as a thought experiment created by William Newcomb of the University of California in 1960 (Wolpert and Benford, 2010), but was first analyzed and published in a Journal of philosophy by Robert Nozick in 1969.

The paradox goes like this: The player of a game is presented with two boxes, one transparent (labeled A) and the other opaque (labeled B). The player is permitted to take the contents of both boxes, or just the opaque box B. Box A contains a visible \$1. The contents of box B, however, are determined by an omnipotent being who can predict everything correctly. The Omnipotent Predictor told the player: "I will make a prediction as to whether you will take just Box B, or both boxes. If I predict both boxes will be taken, I will put nothing in box B. If I predict only box B will be taken, then I will put \$1,000 in Box B." At some point before the start of the game, the God-like being made the prediction and put the corresponding contents (either \$0 or \$1,000) in Box B.

Scenario	God-Like Prediction	The Player's Choice	Payout
1	A and B	A and B	\$1
2	A and B	B only	\$0
3	B only	A and B	\$1001
4	B only	B only	\$1000

The player is aware of all the rules of the game, including the two possible contents of Box B, the fact that its contents are based on the Predictor's prediction, and knowledge of the Predictor's infallibility. The only information withheld from the player is what prediction the Predictor made, and thus what the contents of Box B are.

The problem is called a paradox because two strategies (the Dominance Principle and the Expected Utility Maximization), both intuitively logical, give conflicting answers to the question of what choice maximizes the player's payout (wikipeda.org).

The *Dominance Principle*: This strategy argues that, regardless of what prediction the Predictor has made, taking both boxes yields more money. That is, if the prediction is for both A and B to be taken, then the player's decision becomes a matter of choosing between \$1 (by taking A and B) and \$0 (by taking just B), in which case taking both boxes is obviously preferable. But, even if the prediction is for the player to take only B, then taking both boxes yields \$1001, and taking only B yields only \$1000—taking both boxes is thus better regardless of which prediction has been made.

Newcomb's Paradox

The *Expected Utility Maximization*: This strategy suggests taking only B. By this strategy, we can ignore the possibilities that return $0 and $1001, as they both require that the Predictor has made an incorrect prediction, and the problem states that the Predictor is never wrong. Thus, the choice becomes whether to receive $1 (both boxes) or to receive $1000 (only box B)—so taking only box B is better.

In his 1969 article, Nozick noted that "To almost everyone, it is perfectly clear and obvious what should be done. The difficulty is that these people seem to divide almost evenly on the problem, with large numbers thinking that the opposing half is just being silly."

The crux of the paradox is in the existence of two contradictory arguments, both seemingly correct.

If there is no Free Will, we don't have a choice at all. My future actions are determined by the past predictions of God. Since we don't have the Free Will to make choices, the game does not make sense at all. Without Free Will, we even wonder what is the meaning of *if*. If there is Free Will, then we have choice and our choice will affect God's prediction, meaning, our future action will determine the fate of a past event (reverse causation). Interestingly, Newcomb said that he would just take B; why fight a God-like being? (Wolpert and Benford, 2010).

A similar paradox, called the Idle Argument, goes like this: Suppose that the Omnipotent Predictor has predicted the grade for my future exams. Then I can just relax and sit, doing absolutely nothing at all.

6.4 The Monty Hall Dilemma

The *Monty Hall Dilemma* (MHD) was originally posed in a letter by Steve Selvin to the *American Statistician* in 1975 (Selvin 1975a, 1975b). A well-known statement of the problem was published in Marilyn vos Savant's "Ask Marilyn" column in the magazine *Parade* in 1990 (vos Savant, 1990).

Suppose you're on a game show and you're given the choice of three doors. Behind one door is a car; behind the others are goats. The car and the goats were placed randomly behind the doors before the show. The rules of the game show are as follows: After you have chosen a door, the door remains closed for the time being. The game show host, Monty Hall, who knows what is behind the doors, now has to open one of the two remaining doors, and he will open a door with a goat behind it. After Monty opens a door with a goat, he always offers you a chance to switch to the last, remaining door. Imagine that you chose Door 3 and the host opens Door 1, which has a goat. He then asks you "Do you want to switch to Door Number 2?" Is it to your advantage to change your choice?

Many readers refused to believe that switching is beneficial as Von Savant suggested. Ironically, Herbranson and Schroeder (2010) recently conducted experiments showing that supposedly stupid birds pigeons can make the right decision when facing the Monty Hall Dilemma. They wrote: "A series of experiments investigated whether pigeons (Columba livia), like most humans, would fail to maximize their expected winnings in a version of the MHD. Birds completed multiple trials of a standard MHD, with the three response keys in an operant chamber serving as the three doors and access to mixed grain as the prize. Across experiments, the probability of gaining reinforcement for switching and staying was manipulated, and birds adjusted their probability of switching and staying to approximate the optimal strategy. Replication of the procedure with human participants showed that humans failed to adopt optimal strategies, even with extensive training."

This problem has attracted academic interest because the result is surprising. Variations of the MHD are made by changing the implied assumptions, and the variations can have drastically different consequences.

As the player cannot be certain which of the two remaining unopened doors is the winning door, most people assume that each of these doors has an equal probability of being that door and conclude that switching does not matter. However, the answer may not be correct depending upon the host's behavior. You could increase the probability (p) of winning from $1/3$ to $2/3$ by switching! Here is why. The player, having chosen a door, has a $1/3$ chance of having the car behind the chosen door and a $2/3$ chance that it's behind one of the other doors. It is assumed that when the host opens a door to reveal a goat, as he always does, this action does not give the player any new information about what is behind the door he has chosen, so the probability of there being a car behind a different door remains $2/3$; therefore the probability of a car behind the remaining door must be $2/3$. Switching doors thus wins the car $2/3$ of the time, so the player should always switch. However, the host's behavior might affect the probability and your decision. For instance, if the host is determined to show you a goat; with a choice of two goats (Bill and Nan, say), he shows you Bill with probability q (or he shows you Bill q of the time), then given that you are shown Bill, the probability that you will win by switching doors is $p = 1/(1 + q)$.

The Monty Hall Problem

Now here is an interesting question: If we see Bill, but don't know if the host is always showing Bill or just randomly picking a goat that happens to be Bill this time, how do we determine the probability p? A simple answer would be the Bayesian approach to assume a prior probability and calculate the posterior probability. Now if the host announces honestly that he is doing things differently this time from previous games and will again change the way he shows the goat in the future, i.e., this game is just a one-time event, how do you make the choice or calculate the winning probability p after the switching? Do you use the similarity principle (Section 1.4) to figure out your answer? What is your causal space?

6.5   The Two-Envelope Paradox

The *envelope paradox* dates back to at least 1953, when Belgian mathematician Maurice Kraitchik proposed this puzzle (wikipedia.org):

Two people, equally rich, meet to compare their wallets' contents. Each is ignorant of the contents of both wallets. The game is this: whoever owns the wallet having the least money receives the contents of the wallet of the other player (in the case where the amounts are equal, nothing happens). One of the two men reasons: "Suppose that I have amount A in my wallet. That's the maximum that I could lose. If I win (probability 0.5), the amount that I gain will be more than A. Therefore the game is favorable to me." The other man reasons in exactly the same way. However, by symmetry, the game is fair. Where is the mistake in the reasoning of each player?

In 1989, Barry Nalebuff presented the paradox in the two-envelope form as follows: The player is presented with two indistinguishable envelopes. One envelope contains twice as much as the other. The player chooses an envelope to open, receiving the contents as his reward. The question is, after the player picks an envelope with or without opening it, do you think he should switch to pick the other envelope if he is allowed?

Intuitively, you may say switching or not will make no difference because no information has changed. However, the following reasoning leads to a paradox.

(1) Let A and B be the random variables describing the amounts in the left and right envelopes, respectively, where $B = A/2$ or $B = 2A$ with equal probability.

(2) If the player holding the left envelope makes the decision to swap envelopes, he will take the value $2A$ with probability $1/2$ and the value $A/2$ with probability $1/2$.

(3) The "expected value" the player has by switching is $\frac{1}{2}\left(2A + \frac{A}{2}\right) = 1.25A > A$. So a rational player will select the right envelope expecting a greater reward than from the left envelope. However, since the envelopes are indistinguishable there is no reason to prefer one to the other.

(4) A common criticism to this reasoning is that A is a random variable, standing for different values at different times. Therefore, the expected value calculation, step 3 above, is not correct. In the first term A is the smaller amount while in the second term A is the larger amount. However, this criticism is unsound because the player can open an envelope, see the particular value of A (saying, \$2), and make a decision whether to switch or not. In this case, $A = \$2$ is fixed, not random at all. Therefore, if the player chooses not to switch, he will gain \$2; if he chooses to switch, he will expect to get $0.5(\$1) + 0.5(\$4) = \$2.5$. This analysis is applicable to any value of A. As a result, the player should switch the envelope regardless of what value he sees, which further implies he should make the decision to switch even before he opens the envelope.

The Two Envelope Problem

Can this paradox be resolved from the Bayesian perspective? Blachman et al. (1996) among others pointed out that if the amounts of money in the two envelopes have any proper probability distribution representing the player's prior beliefs about the amounts of money in the two envelopes, then it is impossible that whatever the amount $A = a$ in the first envelope might be, it would be equally likely, according to these prior beliefs, that the second contains $a/2$ or $2a$. In other words, it is impossible for the situation 1 above to hold for all possible values A. Chang (2012) elaborated on this with four different examples. A Bayesian analysis using a nonuniform prior distribution of A will easily solve the problem. However, the use of such a prior distribution will change the original problem with equal probability of having A or $2A$ in either of the envelopes for all possible values of A.

One way to satisfy situation 1 above is this: When you pick an envelope, I will provide you the option to "switch." If you decide to "switch," I will flip a fair coin. If it is heads I will double the money you have in the envelope; otherwise, I will take away half what you have. This is no different from the condition given in the original game. In this case, it is obvious that switching will expectedly increase the value of your money by 25%, but a one-time "switch" is as good as multiple "switches."

6.6   Simpson's Paradox

In probability and statistics, *Simpson's paradox*, introduced by Colin R. Blyth in 1972, points to apparently contradictory results between aggregate data analysis and analyses from data partitioning. Let's look into some examples.

One of the best known real-life examples of Simpson's paradox occurred when the University of California, Berkeley was sued for bias against women who had applied for admission to graduate schools there. The admission rates (44% for men and 35% for women) for the fall of 1973 showed that men applying were more likely than women to be admitted, and the difference was statistically significant (Bickel, Hammel, and O'Connell, 1975). However when examining the individual departments, it was found that most departments had a "small but statistically significant bias in favor of women."

The three authors conclude: "Examination of aggregate data on graduate admissions to the University of California, Berkeley, for Fall 1973 shows a clear but misleading pattern of bias against female applicants. Examination of the disaggregated data reveals few decision-making units that show statistically significant departures from expected frequencies of female admissions, and about as many units appear to favor women as to favor men. If the data are properly pooled, taking into account the autonomy of departmental decision making, thus correcting for the tendency of women to apply to graduate departments that are more difficult for applicants of either sex to enter, there is a small but statistically significant bias in favor of women. The graduate departments that are easier to enter tend to be those that require more mathematics in the undergraduate preparatory curriculum."

However, if we keep hunting for a positive finding, we will eventually find cases of sex bias against women, which might be false findings due to the so-called multiplicity issue to be discussed in Section 6.9.

The second example is about the effects of drugs. Suppose two drugs, A and B, are available for treating a disease. The treatment effect (in terms of response rate) is $520/1500$ for B, which is better than $500/1500$ for treatment A. Thus we will prefer treatment B to A. However, after further looking into the data for males and females separately, we found that the treatment effect in males is $200/500$ with A, better than $380/1000$ with B, while the treatment effect in females is $300/1000$ with A, better than $140/500$ with B. Therefore, whether female or male, we will prefer treatment A to B. Should we take treatment A or B?

	Drug A	Drug B
Male	200/500	380/1000
Female	300/1000	140/500
Total	500/1500	520/1500

The problem can be even more controversial. Suppose when we further look into the subcategories: Young Female and Old Female, the direction of treatment effects switches again, i.e., treatment B has better effect than treatment A in both subcategories, consistent with the treatment effect for the overall population. The question is: What prevents one from partitioning the data into arbitrary subcategories artificially constructed to yield wrong choices of treatments?

	Drug A	Drug B
Young Female	20/200	40/300
Old Female	280/800	100/200
Total	300/1000	140/500

A relevant issue is raised in globalized drug development. As of early 2008, there were 50,629 clinical trials ongoing globally, up by 1.3% over 2007. Clinical trials across multiple regions of the world have become common practice. If the overall drug effect is statistically significant when all regions are combined but very different drug effects were observed in different countries/regions, how should a drug be used in different countries? A second question is: How should the region be defined, as a country, a state, a city, or even something that is not geographically defined?

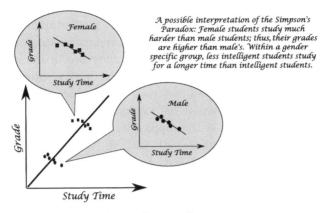

A possible interpretation of the Simpson's Paradox: Female students study much harder than male students; thus, their grades are higher than male's. Within a gender specific group, less intelligent students study for a longer time than intelligent students.

Simpson's Paradox

6.7 Regression to the Mean

The phenomenon of *regression to the mean* was first identified by Sir Francis Galton in the nineteen century. Galton was a half-cousin of Charles Darwin, a geographer, a meteorologist, a tropical explorer, a founder of differential psychology, the inventor of scientific fingerprint identification, a pioneer of statistical correlation and regression, a convinced hereditarian and eugenicist, and a best-selling author.

Galton discovered that sons of very tall fathers tended to be shorter than their fathers and sons of very short fathers tended to be taller than their fathers. A similar phenomenon is noticed: A class of students takes two editions of the same test on two successive days; it has frequently been observed that the worst performers on the first day will tend to improve their scores on the second day, and the best performers on the first day will tend to do worse on the second day. This phenomenon is called "regression to the mean," and is explained as follows. Exam scores are a combination of skill and luck. The subset of students scoring above average would be composed of those who were skilled and did not have especially bad luck, together with those who were unskilled but were extremely lucky. On a retest of this subset, the unskilled will be unlikely to repeat their lucky performance, while the skilled will be unlikely to have bad luck again. In other words, their scores will likely go back (regress) to values close to their mean scores.

The phenomenon of regression to the mean holds for almost all scientific observations. Thus, many phenomena tend to be attributed to the wrong causes when regression to the mean is not taken into account. What follows are some real-life examples.

The calculation and interpretation of "improvement scores" on standardized educational tests in Massachusetts provides a good example of the regression fallacy. In 1999, schools were given improvement goals. For each school the Department of Education tabulated the difference in the average score achieved by students in 1999 and in 2000. It was quickly noted that most of the worst-performing schools had met their goals, which the Department of Education took as confirmation of the soundness of their policies. However, it was also noted that many of the supposedly best schools in the Commonwealth, such as Brookline High School (with 18 National Merit Scholarship finalists), were declared to have failed. As in many cases involving statistics and public policy, the issue was debated, but "improvement scores" were not announced in subsequent years, and the findings appear to be a case of regression to the mean (wikipedia.org).

The psychologist Daniel Kahneman, winner of the 2002 Nobel Prize in Economics, pointed out that because we tend to reward others when they do well and punish them when they do badly, and because there is regression to the mean, it is part of the human condition that we are statistically punished for rewarding others and rewarded for punishing them (Daniel Kahneman's autobiography at www.nobelprize.org).

Regression to the mean is a common phenomenon in clinical trials. Because we only include patients who meet certain criteria, e.g., hemoglobin levels lower than 10, among patients who enter the study, some are accidently lower than 10 and will regress to the mean late on in the trial. Therefore, the overall treatment effect usually includes the treatment effect due to the drug, the placebo effect, and a part due to regression to the mean.

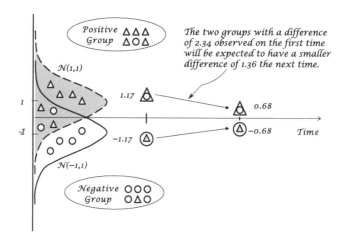

Regression to the Means of the Two Large Normal Samples

6.8 Causation, Association, Correlation, and Confounding

We are going to discuss the interesting relationships among causation, association, correlation, and confounder. We use the following 20 sample data points measured on ten subjects in a hypothetical experiment, in which 5 are treated with placebo ($X = 0$) and 5 are treated with the test drug ($X = 1$). Two variables are measured, the response Y and a biomarker Z. For clarity, assume there no any other hidden variables that affect the response; thus, no randomness is involved. The population is simply replications of the set of these 10 points.

Response and Biomarker for 10 Subjects

	Placebo					Test Drug				
Response Y	1	2	3	4	5	1	2	3	4	5
Biomarker Z	1	2	3	4	5	3	4	5	6	7

Because there is no randomness involved, the relationship among response Y, variable Z, and treatment group X can be perfectly modeled using a linear equation,

$$Y = Z - 2X.$$

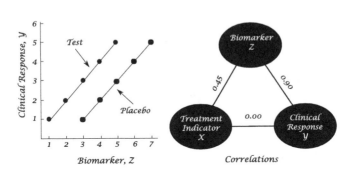

Complexity of Biomarker

How to interpret the result is somewhat tricky, being dependent on the biomarker's characteristics and when it is measured, as we now elaborate.

(1) If the biomarker Z is a baseline measurement, i.e., measured before treatment, for Z between 3 and 5, we see that the test drug has an effect of 2 more units than the placebo in response. For $Z = 1$ and 2 we only have data for the test group, and for $Z = 6$ and 7 we only have data for the placebo group. When we use a regression model in statistics, we will get the same equation but usually it would be applied to the whole range of Z, from 1 to 7. In this scenario Z is called a confounder (Section 3.7), because without Z

in the model we would conclude differently, i.e., that the treatment X has no effect on response Y.

(2) If Z is a post-baseline measurement after treatment and Z (such as with a DNA marker) is not affected by treatment X, then the situation is virtually the same as Z being the baseline measurement; but if Z can be modified by X (as in the case of an RNA marker), then things become complicated. On the one hand, if the difference in Z between the two groups is purely due to X (no difference in Z between the two groups at the baseline), then there is no treatment effect of X on the response Y. In other words, anyone who has the same baseline value Z will have the same response Y regardless of the treatment X they taken. What the test drug does is to reduce the biomarker Z. In this scenario, Z should not be in the model. Otherwise, one would mistakenly conclude that the test drug will affect the response Y. However, if the difference in Z between the two groups is partially at the baseline and partially due to treatment X, then the equation is difficult to interpret. If this is the case, we should use the model that includes both the baseline Z and the post-baseline Z as separate factors.

(3) The third interesting point is that, like friendship, the correlation is not transitive: we see that the correlation between X and Z is 0.45 in this example, and that between Z and Y it is 0.9, but this does not lead to a positive correlation between X and Y—the correlation between X and Y is zero. Recall from Section 4.1 that a zero Pearson's correlation coefficient does not mean there is no association but means only no positive or negative linear association. But this statement seems not to hold at the moment because we have zero correlation and a perfect linear relationship between X and Y. This appears somewhat paradoxical because we are using a different number of variables in the two situations. Here we also see that association does not necessarily mean causation, as discussed in different scenarios above.

6.9   Multiple Testing

Scientists conduct experiments because they believe there are causal relationships in the world. They propose hypotheses and hope to prove or disprove them through hypothesis testing. However, we have seen many controversies and paradoxes surrounding scientific inference (Chang 2012). We are going to discuss a very challenging issue in the quantification of scientific evidence: multiplicity. The word *multiplicity* refers to the statistical phenomenon of error inflation in falsely rejecting the null hypothesis when multiple tests are performed without the so-called multiplicity adjustment to the level of significance, α. While this technical definition may pose some initial difficulty to the nonspecialist, an understanding can be achieved based on the examples given below. Multiplicity exists in every aspect of science and daily life. Just as the black hole that attracts everything, multiplicity is the horizon where all different statistical paradigms merge.

In a national lottery a few years ago, there was one and only one winning ticket. Ten million people each bought one ticket, so each person had a very small chance of winning—1 out of 10 million. A Mr. Brown was the lucky winner, but he also was the object of a lawsuit charging him with conspiracy. The prosecution argued that the chances of Mr. Brown winning the lottery were so low that, practically speaking, it could not happen unless there was a conspiracy. However, Mr. Brown defended: "There must be a winner whoever he/she might be, regardless of any conspiracy theory." This paradox is a typical multiplicity problem. The multiplicity emerges when we apply the conspiracy theory (the null hypothesis is that there is no conspiracy) to each of the ten million lottery buyers.

Suppose you tell a friend that you have a biased coin (heads are more likely than tails) and he wants to prove it with an error rate that is no more than 5%, i.e., at level of significance of 0.05. So you set the null hypothesis: the probability of heads is equal to 0.5, and the alternative hypothesis is that the probability of heads is larger than 0.5. You believe that the bias/difference in probability is large and an experiment consisting of 100 coin-flips (Bernoulli trials) will provide sufficient power to detect the bias. Based on a binomial distribution, you know that if you can show 59 or more heads out of 100 trials, then you have proved the coin is biased in favor of heads. However, you think you may only need 50 trials and if you can show 32 heads out of 50 trials, you have proved the coin is biased. Suddenly, you have an exciting new idea: do 50 trials first, and if 32 of 50 are heads, the coin is proved biased; if not, you will do another 50 trials and if there are 59 heads out of 100 trials, you will claim that the coin is biased. In this way, you have two chances to prove your theory, and at the same time, you will also have two chances to make false claims: the first one at the analysis on 50 trials and the second one at 100 trials. Therefore, to control the type-I error, you have to adjust the hurdles, α, downward to smaller values. This is essentially the idea of controlling type-I error in an adaptive design.

Now suppose you are diagnosed with a rare disease, and the doctor provides you with two options: Drug A or Drug B; each has been tested on sets of 100 patients to see if it is better than a control treatment with a known 50% cure rate. In the first clinical trial, 59 are cured out of 100 patients treated with Drug A, which is statistically significant at the one-sided 5% level. In the second clinical trial, 60 patients out of 100 are cured with Drug B. But this is not even statistically significant, because in the second trial there is an interim analysis of the first 50 patients where 30 of them were found cured, and, as we explained in the previous coin-flipping experiments, the hurdles for rejecting the null hypothesis are higher when there are multiple analyses. Here is the dilemma: Drug B shows a better cure rate of 60% versus Drug A at 59%. But drug A is statistically better than the control and Drug B has not achieved statistical significance in comparison with the control treatment. Which drug do you take? On one hand, you may decide to take drug A since drug B does not even achieve statistical significance, implying that drug B appearing better than the control may just be random chance. On other hand, you may conclude that with or without the interim analysis of 50 patients, the chemical structure of the drug will not change. In this case you can ignore the interim analysis; a cure rate of 60% is better than 59%, so you should take Drug B. We should know that when there are many interim analyses, a cure rate of 90% may not be statistically significant because the α for each test is adjusted to a very low value.

Patient's Dilemma

Here, we can see two different concepts of the effectiveness of a drug. One is the physical properties of the test compound, which will *not* change as the hypothesis test procedure changes (e.g., one test versus two tests). The other is the statistical property, which will change since it reflects an aggregated attribute of a group of similar things, such a property will depend on the choice of causal space (a collection of similar "experiments") as we have discussed earlier.

Let's introduce the example given by Vidakovic (2008): Suppose a burglary has been committed in a town, and 10,000 men in the town have their fingerprints compared to a sample from the crime. One of these men has a matching fingerprint, and at his trial it is testified that the probability that two fingerprint profiles match by chance is only 1 in 20,000. However, since 10,000 men had fingerprints taken, there were 10,000 opportunities to find a match by chance; the probability of at least one fingerprint match is 39%, which is considerably more than 1 in 20,000.

Now suppose Interpol were going to run a fingerprint check according to the alphabetic order of last names, starting with the letter A, and that the check is to be stopped if and when a match is found. Suppose that the suspect's fingerprints were the very first checked (his last name happened to be Aadland) and that they matched the crime sample. Thus, the probability is 1/20,000. But the suspect could argue that if the agency's check started with last names beginning with Z and even if the agency wanted to check everyone, they would have identified all 0.36 million fingerprint matches in the world population of 7.20 billion before checking on him. Therefore, fingerprints (and, similarly, DNA matching) can only be used for excluding suspects, but not for convictions. Do you agree with this argument?

The multiplicity issue does not necessarily explicitly involve multiple hypothesis tests. It can arise from model fitting by trying multiple models, or from analyses performed on the same data by different researchers. Multiplicity issues constantly appear in publication, even in prestigious journals. Alec Beall and Jessica Tracy (2013) published an article in *Psychological Science* entitled: "Women Are More Likely to Wear Red or Pink at Peak Fertility."

"Although females of many species closely related to humans signal their fertile window in an observable manner, often involving red or pink coloration, no such display has been found for humans. Building on evidence that men are sexually attracted to women wearing or surrounded by red, we tested whether women show a behavioral tendency toward wearing reddish clothing when at peak fertility. Across two samples (N = 124), women at high conception risk were more than 3 times more likely to wear a red or pink shirt than were women at low conception risk, and 77% of women who wore red or pink were found to be at high, rather than low, risk. Conception risk had no effect on the prevalence of any other shirt color. Our results thus suggest that red and pink adornment in women is reliably associated with fertility and that female ovulation, long assumed to be hidden, is associated with a salient visual cue."

Professor of Statistics Andrew Gelman from Columbia University pointed out several shortcomings of the study, including representativeness and bias. He also raised potential multiplicity issues and rejected the claim that women are three times more likely to wear red or pink when they are most fertile. "No, probably not. But here's how hardworking researchers, prestigious scientific journals, and gullible journalists have been fooled into believing so."

I am not sure I completely agree with Professor Gelman's conclusion, but I very much agree with his concern for multiplicity problems in general in publications, and I believe further validation of the "Pink Shirt Finding" is necessary to make any wide conclusion. I also very much appreciate Gelman's skill in critical review here, not particularly in holding against the "Pink Shirt" paper but in putting on display the good habit of reading published articles carefully. It is essential for all scientists to adopt such a critical attitude in reviewing articles.

It is interesting to anticipate the trend that, as data sources gradually become public, more researchers will perform different analyses on the same data. How can we control the error rate if the first research has already used up the error rate allowed (α)? On the other hand, the conclusions from meta-analyses (post hoc analyses using combined data from similar experiments) are usually difficult to disprove since (a) most, if not all, available data have been used, and (b) with such a large sample size the conclusion is difficult to change, even if we have new data later. The question is: Should we use the conclusions drawn from a meta-analysis and disregard a previous conclusion (whether is positive or negative) from a subset of studies, each with a smaller sample size? How can we make use of a data resource multiple times for many different questions but without inflating α, or with only a limited inflation? We will further discuss the multiplicity issue under the title of Publication Bias in Section 7.20.

In current practice, the type-I error rate is controlled within an experiment. Thus, different pharmaceutical companies test chemical compounds every day from the same or similar compound libraries, a majority of them being ineffective for their intended purpose. Because the tests are performed by different companies, no multiplicity adjustments are applied. Therefore, the false positive rate in such screening can be high. Similarly, if an ineffective compound is tested 10 times using 10 clinical trials with 10 patients each versus one trial with 100 patients, the error rate will increase from 5% to 40%.

Now here is a question for you to think about: Should one consider one's lifetime as a big experiment and control the "type-I error"?

There are many ways to control a false positive rate. For instance, we apply a smaller α to even-th experiments or analyses and a larger α to odd-th experiments or analyses with a larger α, but overall type-I error is controlled. This is similar to adaptive design, where one can distribute α at the interim and the final analyses differently.

Both frequentist and Bayesian statisticians address the false positive error rate, but their approaches are different (Chang 2011).

6.10   Exploratory and Confirmatory Studies

Most of us try different things in our lifetimes to discover our interests and what we are truly good at. These are *exploratory studies*. We then test these initial findings, for example, and choose our professions based on the results. In this case, one's daily routine in the workplace can be considered a confirmatory study. It can turn out to be enjoyable and successful, or it might not be. The outcome may also be dependent on other factors or simply chance. If not successful we can try something else, starting an exploratory study again.

Likewise, to solve a complex research problem, an investigation is usually started with exploratory studies and followed by one or more confirmatory studies. Consider pharmaceutical research. The first exploratory study can be a brainstorming session or qualitative study. Late stage exploratory studies can include experimental design with exploratory questions such as: Does the test drug generate certain desirable biological activity? What is the dose–response curve? What is the best dose for treating humans as extrapolated from animal studies? A *confirmatory study* is often a hypothesis-based study to confirm initial findings from previous exploratory studies. A confirmatory study is often larger and more costly, and requires a longer time to conduct than an exploratory study. For example, in drug development, the hypothesis may be: the selected dose of the test drug is efficacious and safe for the intended patients.

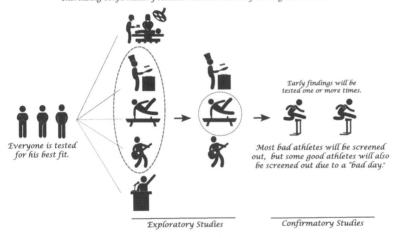

A Key to Success Is to Find Balance between Exploratory and Confirmatory Studies: Increasing exploratory studies will increase the false positive error; Increasing confirmatory studies will increase the false negative error.

Everyone is tested for his best fit.

Early findings will be tested one or more times.

Most bad athletes will be screened out, but some good athletes will also be screened out due to a "bad day."

Exploratory Studies   *Confirmatory Studies*

Exploratory and Confirmatory Studies

Exploratory and confirmatory studies are not only different in their objectives and study designs, but also different in data analysis. An *exploratory analysis* often uses descriptive statistics, encourages deeper understanding of mechanisms or processes, requires judgment and artistry, and usually does not provide definitive answers. The open-minded nature of exploratory analysis can be described by three characteristics: mild, flexible, and post hoc (Goeman and Solari, 2011). The inferential procedure is mild because the level of significance doesn't have to be specified before the analysis. For example, the top five promising drug candidates, based on their ranking in effect sizes or *p*-values, will be selected for the next study, hoping that a majority of the false rejections will be removed then. The procedure is flexible because it does not require prescribing hypotheses, but leaves the choice of which hypotheses to follow up to the researcher. Furthermore, the procedure is post hoc, because it allows the research to determine the hypotheses after the researcher reviews data or the analysis results. Confirmatory analysis, in contrast, is rigorous in the control of false positive errors.

Taking drug development as an example, thousands of chemical compounds are screened using in vitro experiments (e.g., in diseased tissues from animals) every day for biological effects. We can imagine there are many false positive findings (FPF). How do we reduce the FPFs? Those compounds that show biological effects will be selected for further in vivo tests (in animals) and a smaller number of compounds that continue showing the desirable activities will be further tested in animals to determine the optimal dose range for safety and efficacy. Next, the determined dose range in animals will be extrapolated to the dose range for human by so-called biological scaling (see Section 7.11). After all these steps, the tests of the compound in clinical trials (experiments in humans) start, initially with a very conservative dose. There are three phases in clinical trials. Phase-1 trials are designed to test the safety of the drug in humans, Phase-2 trials are used to further confirm the safety and determine the dose response curve, and the best suitable dose will then be selected to further test in Phase-3 studies. All the experiments before Phase-3 are conventionally called exploratory studies, while the large, costly Phase-3 clinical trials are called confirmatory studies. At any step of the process a failure can occur (either true or false negative claims). To reduce the cost and risks to the patients, the sample size usually increases gradually from Phase-1 to Phase-3 trials. If a drug is proved to be effective and safe in Phase-3 studies, the drug will be approved by the authority (the FDA in the US) for marketing.

The current trial, whether exploratory or confirmatory in nature, often serves as a study to confirm the findings from the previous study. Such a confirmation is weak with exploratory studies and strong or rigorous with confirmatory studies.

In determining the overall research strategy, it is critical to balance between the exploratory and confirmatory studies, because differences in experiment type and size imply differences in cost and the probability of false positive and false negative findings. To determine the balance point, it is helpful to understand the so-called union and intersection hypothesis tests.

Suppose we test the following two claims (alternative hypotheses) in an experiment:

$$H_1 : \text{Drug } X \text{ is effective for treating disease } D,$$

and

$$H_2 : \text{Drug } Y \text{ is effective for treating disease } D.$$

In an exploratory analysis, the significance level for each hypothesis test, α, is usually unchanged; thus, the probability of making a false positive claim is more than α, i.e., the type-I error inflation. This is because there are two chances to making a false positive claim, one for each claim. The more hypotheses we perform in exploratory analyses, the higher the type-I error inflation will be. Such an error inflation can be very high when we test, for example, a thousand compounds and/or genetic markers in the drug development process. On the other hand, in a confirmatory analysis, the α for each test is adjusted (e.g., to $\alpha/2$ when using the so-called *Bonferroni correction*) so that the type-I error will not be inflated.

The above analysis is called a *union test*. The probability of rejecting the union test is

$$\Pr(H_1 \text{ or } H_2) = \Pr(H_1) + \Pr(H_2) - \Pr(H_1 \text{ and } H_2).$$

Because in a confirmatory analysis, we reduce alpha to $\alpha/2$ for each test, i.e., $\Pr(H_1) \leq \alpha/2$ and $\Pr(H_2) \leq \alpha/2$, when both H_1 and H_2 are true, we ensure $\Pr(H_1 \text{ or } H_2) \leq \alpha/2 + \alpha/2 = \alpha$.

In contrast, in another kind of test, the *intersection test*, the hypotheses H_A and H_B appear in sequence, the later one providing a retest or confirmation of the previous one. For example, suppose, in an experiment, we are interested in two hypotheses:

$$H_A : \text{Drug } X \text{ increases hemoglobin at the third week,}$$

and

$$H_B : \text{Drug } X \text{ increases hemoglobin at the fifth week.}$$

To claim that drug X is effective, we must prove H_A and H_B. In this case, if we apply the significance level, α, to each test, the type-I error will not exceed α. This is because in order to test H_B, we have to prove H_A first. Such a combination of tests is an intersection test. The intersection test will usually reduce the probability for making the positive claim because there are multiple hurdles that need to be cleared. When H_A and H_B are independent, the probability of rejection in an intersection test is the joint probability

$$\Pr(H_A \text{ and } H_B) = \Pr(H_A) \Pr(H_B),$$

which is smaller than both $\Pr(H_A)$ and $\Pr(H_B)$. Thus, the probability (power) of discovering a true positive drug will decrease and the probability of a false negative increases. For example, for two independent hypothesis tests (more precisely two independent test statistics) each with 90% power, the probability for the candidate to pass the two tests is $0.9 \times 0.9 = 0.81$ and probability of false negative findings is $1 - 0.81 = 0.19$. If there are five hypotheses in an intersection test, the power is $0.9^5 = 0.59$.

We should be aware that in current practice, the type-I error is controlled within an experiment for a confirmatory study, which is called experiment-wise or family-wise error control. Therefore, whether the previous two hypothesis tests (H_1 and H_2) are performed within a single experiment or in two separate experiments will imply different probabilities of a false positive finding. The former has a smaller probability of false positive findings because the α needs to be adjusted to a smaller value, whereas the latter has a larger probability of false positive findings because the α need not be adjusted for either of the two hypotheses simply because they are tested in two different experiments. Such a practice appears to be inconsistent and may surprise you. We will discuss further the controversies of multiplicity in publication bias in Chapter 7.

To understand why it is critical to balance between exploratory and confirmatory studies, it would be helpful for readers to go through the following thought experiment:

Imagine you are asked to find, among all of the different types of coins in the world (e.g., U.S. coins are the penny, nickel, dime, quarter, dollar), which coin types are biased, i.e., tosses resulting in heads and tails have different probabilities. You are asked to do this as quickly as possible with the smallest number of coins. There are lots of different types of coins, maybe 100,000, that have been made in world history. You may expect there are a small number of coins that are biased, but you are not sure. You can only perform coin-flip experiments to find/prove a biased coin, but remember that there are experimental errors. Also, there are penalties when errors are made. What is your research strategy?

6.11   Probability and Statistics Revisited

"It is unanimously agreed that statistics depends somehow on probability. But, as to what probability is and how it is connected with statistics, there has seldom been such complete disagreement and breakdown of communication since the Tower of Babel.... Doubtless, much of the disagreement is merely terminological and would disappear under sufficiently sharp analysis" (Savage, 1954, p.2).

When in daily conversation we discuss "the probability of winning the lottery" or "the probability of rain tomorrow," we don't care about the precise definition of probability. However, for academic discussion and scientific research, it is important to have a precise definition. Such a definition should be consistent across different statistical paradigms. If we have such a consistent definition, the difference in the calculation of a probability is secondary, and usually straightforward.

As we discussed in Chapter 4, there are two problems with the definition of probability on the basis of repeated identical experiments: (the first is that) "the same experiments" cannot be exactly the same, since if they were exactly the same then the outcomes of all the experiments would be the same. Hence, the second problem is that the same experiments mean actually similar experiments. But just what makes up "similarity" has never been made explicitly clear, and thus, people calculate probabilities differently because they implicitly use different definitions of what it means for the two experiments to be "similar."

Let's discuss the concept of probability further with examples, including the one we discussed earlier.

(1) If I toss a Chinese coin once, what is the probability the coin comes up heads? It will be either heads or tails; the result is fixed but unknown.

Definition: Flip the same Chinese coin under similar conditions many (or infinitely many) times, the proportion of heads that we perceive is the probability of heads.

A frequentist statistician may proceed like this: "Since I don't have the experimental data, I have to flip the Chinese coin and calculate the proportion of heads."

But the Bayesian calculation is more like this: "Since U.S. coins are similar to Chinese coins, I use the probability for flipping U.S. coins for the estimation until the data of flipping Chinese coins really starts accumulating. More precisely, the combination of my prior knowledge about U.S. coins and the data from the experiment on Chinese coins produce the probability of interest."

Both frequentist and Bayesian statisticians will come to the same conclusion if the experiment could be conducted infinitely times! Therefore, what they are ultimately looking for (i.e., the definition of probability) is the same, i.e., the proportion of favorable outcomes among all possible outcomes, but the calculations of the probability are different.

(2) What is the probability of the Boston Celtics winning the basketball game against the New York Knicks last night? In this case, the event is not random, it has already happened, so there seems to be no probability involved. From a knowledge perspective, to those who knew the game's result there is no probability involved. But for those who don't know the outcome, their knowledge about the two possible outcomes can be characterized by the following probability.

Definition: If the Celtics play many ("infinitely many") times against the Knicks, the proportion of Boston win *is* the probability.

A frequentist statistician may calculate the probability something like this: He first defines what the "same experiment" is, for which he may choose all the games Celtics have played and will play against Knicks. He then uses the data available so far to calculate the probability, i.e., the proportion of Celtics wins among all games played against the Knicks, since the future data are not available.

A Bayesian statistician may calculate the probability in the following way. All the games that the Celtics have played against other teams provide the information about the skills of the Celtics team. This information can be considered as a prior. The games the Celtics played against Knicks are the "experimental data" to date. The prior and the data are combined using the so-called Bayes law to obtain the (posterior) probability.

As the two teams play more games, the calculated probabilities of winning using frequentist and Bayesian methods will eventually be the same, as defined.

(3) What is the probability of rain tomorrow? It is a one-time event, and there is nothing random about it.

Definition: Considering that the same (more precisely similar) weather conditions as tonight appeared many millions of nights in the past and, are expected, to occur in the next 1 billion years, the proportion of subsequent rainy days from the past and future is the probability.

The frequentists may calculate the probability like this: Look at the historical record of weather report. Collect all the nights with similar weather conditions as tonight and get the proportion of rainy days on the next days— the probability.

Bayesian statisticians may calculate the probability as follows: Use all the knowledge that is considered relevant. Such prior knowledge will combine the new data collected, as a frequentist would do, to form the posterior probability (updated knowledge) over time.

Ultimately, as the "same experiment" is repeated over time and the data are accumulated, the two calculated probabilities will approach a common value. This common value is the proportion of rainy days among all the days under consideration, and is the true probability that the statisticians of the two different philosophies try to estimate (calculate).

From these examples and the ultimate things that the frequentist and Bayesian statisticians are looking for in probability, we see clearly that their apparent difference in the notion of probability can be unified under the umbrella of collection or repetitions of "same experiments." In other words, the notion of probability is constructed on the more fundamental principle—the similarity principle (Section 1.4). The collection of the "same experiments" is actually the causal space (Sections 1.4 and 4.7). The ultimate causal space can be the same for frequentism and Bayesianism, but for the purpose of calculating probability before we have the ultimate set of data, different statisticians may use temporarily different causal spaces. Over time such calculation of probability becomes more accurate and precise. For this reason, the value of probability can be interpreted as a measure of the status of individual knowledge about a hypothesis. Given all the data they wish to have, the frequentist and Bayesian statisticians will provide the identical value for the probability unless they implicitly use different causal spaces.

In the previous examples, we didn't qualify what "similar conditions" for the same experiment means or what "relevant data" means for frequentists and Bayesians. We called the set of "similar conditions" the causal space in Chapter 4 to reflect the notion that this set of similar conditions is the basis of causal reasoning.

Whether you call yourself a Bayesian or frequentist statistician, you cannot completely avoid the Bayesian notion of probability in your daily life and in your scientific research. We often use Bayesian reasoning without realizing it. For instance, I suspect all of us (Bayesian or frequentist) will answer both the winning team question and the rainy day question from a Bayesian perspective. And you (and I) would likely give a Bayesian answer to the coin-tossing problem before having experimental data to go on. There are many more examples. We can even say that every conscious action we take is a utilization of prior relevant experiences or knowledge so as to be able to probabilistically predict the consequences. Such prior knowledge or experiences are not considered as current "experimental data."

Probability as Status of Knowledge

An event (heads or tails in flipping a coin) will either occur or not. Therefore, knowledge of an event's occurrence is either correct or not. When describing knowledge in probability or making a probabilistic estimate, we implicitly apply the similarity principle: similar situations will likely result in the same outcome. Thus, the results from these similar situations can be pooled together to form (implicitly) a probability distribution of the outcome. However, the method of pooling can be different for different statistical paradigms and even varies from person to person. In the Bayesian paradigm, the set of similar situations is often vaguely or implicitly defined; thus, different researchers might come up with very different values for a probability.

In the *Bayesian paradigm*, the subjective prior is often formulated implicitly based on individual experience, knowledge, or belief. Well, one can argue that a belief comes from experience, even though it might be impossible to explicitly specify the experience.

A statistical principle, like other scientific principles, has to be evaluated for its repetitive (not one-time) performance, where the evaluation has to be based on a certain evaluation matrix or evaluation criteria. The choice of the matrix can vary from time to time, but it has to match the purpose of using the principle. We have learned that there are different statistical paradigms that use different causal spaces and different evaluation criteria. Therefore, with the same data, different paradigms can give different or even conflicting answers to the same problem. The decision on the paradigm to be used, in what circumstance, should be based on the research objectives.

In the *frequentist paradigm*, the same data can lead to different conclusions depending on whether the data are obtained from a single experiment or multiple experiments. This is because the type-I error rate is controlled within an experiment (Section 6.9). In contrast, in the Bayesian paradigm (the Bayes theorem), the final conclusion (posterior distribution) is expected to be the same whether the set of data is used piece-wise (experiment by experiment) or is all used at one time. We can even switch the sequence of the experiments and the final posterior distribution will be the same. Such a consistency in theory can be considered a good thing since the same set of data should contain the same amount of information and should lead to the same conclusion. However, the necessity of maintaining consistency can be challenged philosophically. First, such a consistency is ensured only when the same model is used in the sequential analyses. However, there is no reason for us to just update the model parameters but never update the model itself based on new data. Second, we need to decide whether the term *information* should be dependent on the "sender (source)" or the "receiver (destination)." If the amount of information is determined by the receiver, then the order in which the pieces of information (data) are received is often critical. For instance, teaching an elementary student multiplication before addition would lead to a completely different result from that when teaching the student addition before multiplication. Likewise, if we record two lectures and replay them in two possible different orders, the resulting effects to an audience can be very different. Furthermore, we know that biological sensibilities are most evident on a log-scale or on the basis of percent change, which is the reason we often measure the volume of voice and intensity of light on a log-scale. If a person lifts a 10kg weight and then a 11kg weight, he may be able to tell the difference. However, if the same person lifts the 51kg weight first and then the 50kg weight, he may not notice the difference. This is because in the two scenarios the percentage changes in weight are different, $1/10$ (10%) versus $1/11$ (9.1%). For the same reason, when we are older we feel that a 5-year time period is shorter than the same 5-year time period when we were younger. From these examples we can see that the amount of information from the "sender" seems unchanged, but the receiver entertains different amounts of information. Importantly, this is not because the receiver does not pay attention to the information, but instead is due to an intrinsic trait of all humans. Therefore, from the learning viewpoint it seems more appropriate to measure the amount of information from the receiver's perspective.

Practically, when we collect and analyze data through a sequence of experiments, the correlation between the data may not be as clear as when they are collected in a single experiment and analyzed. Therefore, it is likely that the two approaches will use different correlation structures for the same set of data and lead to different conclusions. For this reason, the number of the experiments and the order in which the experiments are conducted are practically important.

In Section 7.21, we will discuss how a poem can magically produce information resonance. Indeed, from time to time a tiny bit of information (from the sender's point of view) can be so enlightening that a chain of long-unsolved puzzles is unraveled instantly.

Chapter 7

Case Studies

7.1 Social Genius of Animals

Elephants are widely acknowledged to be high on the list of species displaying significant social intelligence and compassion. Several experiments have been conducted to study their cooperative nature. One such experiment is described at www.arstechnica.com:

The Cooperative Nature of Elephants

There are three key steps in the experiment: (1) Elephants are trained to pull a rope attached to a food bowl using a simple apparatus. The idea is: pull the rope, get food. (2) Here a more complex apparatus is used that requires two elephants to synchronize their pulling to attain the food reward. Because the elephants are released at the same time, most likely they'll pull the ropes at about the same time. So the fact that two elephants can perform the job doesn't mean they knowingly cooperate. It could be just the "see the rope, pull the rope, get food" knowledge obtained from the first step.

(3) Using the apparatus as in Step 2, the elephants are released at different times. In order to get the food, the elephant released earlier has to wait for the second one to pull its rope (at the same time) to get the food. If they can do that, cooperation certainly occurs; otherwise, no cooperation can be claimed.

The experimental results from the last stage were presented as follows:

> One elephant was released into the lane with a head start, and the second elephant was released five seconds later. In order to succeed at this task, the first elephant to be released had to recognize the need for a partner and had to wait to pull until the second elephant was able to help. Once the pair had successfully completed the task three times in a row, the head start was increased to 10 seconds, then to 15, then all the way to 25 seconds.

> The total number of trials that the pair required to reach the 25-second mark was then tallied. All the elephant pairs completed this stage in 30 or fewer trials (A "perfect" score, if the elephants had made no mistakes at all, would have been 15 trials). Individual elephants made between three and six errors during this stage; in a study with a similar setup, chimpanzees made up to 28 errors.

> Interestingly, some of the elephants adopted slightly different strategies from the rest. For example, instead of pulling, one young female simply put her foot on her end of the rope. Her partner then arrived and pulled, retrieving the food without the young female doing any extra work at all. Another elephant waited for his partner at the release point, rather than at the apparatus. Alternative strategies such as these are often cited as examples of creativity and advanced intelligence.

7.2   Mendel's Genetics Experiments

Gregor Mendel, is known as the "father of modern genetics." Between 1856 and 1863 Mendel cultivated and tested some 29,000 pea plants (i.e., *Pisum sativum*). Mendel's findings allowed other scientists to predict the expression of traits on the basis of mathematical probabilities. Mendelism formed the core of classical genetics after being integrated with Thomas Hunt Morgan's chromosomal theory of inheritance circa 1915.

Mendel discovered that by crossing white flowered and purple flowered plants, the offspring were not a mix of the two. Instead, the offspring had a ratio of 1:3 between purple and white flowers.

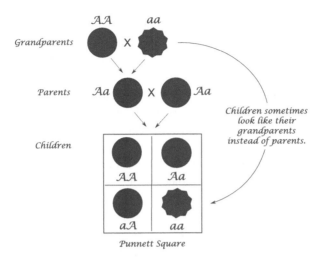

Punnett Square

Phenotype is Determined by the Dominant Gene A

This can be explained as follows. An individual possesses two *alleles* (alternative forms for the same gene) for each trait; one allele is given by the female parent and the other by the male parent. They are passed on when an individual matures and produces gametes: egg and sperm. When gametes form, the paired alleles separate randomly so that each gamete receives a copy of one of the two alleles. The presence of an allele doesn't promise that the trait will be expressed in the individual that possesses it. In *heterozygous* individuals the only allele that is expressed is the dominant. The recessive allele is present but its expression (appearance or phenotype) is hidden. For Mendel's experiment, genes can be paired in four ways: *AA*, *Aa*, *aA*, and *aa* with equal probability. The capital A represents the dominant factor and lowercase a represents the recessive. In heterozygous pairs (*Aa* or *aA*), only gene *A* is expressed, i.e., the same as the homozygous pairs, *AA*. The only pair with a different expression is the homozygous pair *aa*. In other words, the ratio of offspring with the two colors is 1:3.

Mendel's findings are summarized in two laws: the Law of Segregation and the Law of Independent Assortment.

The *Law of Segregation* (The "First Law"): When any individual produces gametes, the copies of a gene separate, so that each gamete receives only one copy. A gamete will receive one allele or the other.

The *Law of Independent Assortment* (The "Second Law"): Alleles of different genes assort independently of one another during gamete formation.

Some decades after Mendel's groundbreaking work, which was generally well accepted, there surfaced criticisms of its accuracy: Statistical analyses suggested that Mendels's data appeared to be "too good," considering randomness in the experimental results. It could be the case that Mendel reported only the best subset of his data, as he mentioned in his paper that it was from a subset of his experiments.

We now know that base pairs are the low-level alphabet of *DNA* instructions, encoding instructions for the creation of a particular amino acid. The base pairs of two nucleic acid bases then chemically bond to each other, forming a 3D double helix ladder of base pairs. There are only four different bases that appear in DNA, i.e., adenine (A), guanine (G), cytosine (C), and thymine (T). The rules for base pairings are simply A pairs with T; G pairs with C. Thus, anyone of the following four base pair configurations composes a single piece of information in the DNA molecule: $A \sim T$, $T \sim A$, $G \sim C$, $C \sim G$.

A *chromosome* is an organized structure of DNA and protein in cells. It is a single piece of coiled DNA containing many genes, regulatory elements, and other nucleotide sequences. Chromosomes also contain DNA-bound proteins, which serve to package the DNA and control its functions.

A *codon* is a template for the production of a particular amino acid or a sequence termination codon. There are sixty-four different codons or different ways to order four different bases in three different locations. But, there are only twenty amino acids for which DNA codes. This is because there are often several different codons that produce the same amino acid.

Producing amino acids is not the end product of DNA instructions. Instead, DNA acts on cells by providing the information necessary to manufacture *polypeptides*, *proteins*, and nontranslated *RNA* (tRNA and rRNA) molecules, each of which carries out various tasks in the development of the organism. Proteins are complex organic molecules that are made up of many amino acids. Polypeptides are protein fragments.

Exchange of genetic material in sexual reproduction takes place through recombination. The DNA from both parents is recombined to produce an entirely new DNA molecule for the child. For human gametes, with 23 pairs of chromosomes, there are 2^{23} or 8,388,608 possible and equally likely combinations.

7.3   Pavlov's Dogs, Skinner's Box

Ivan Pavlov, a Russian psychologist, was awarded the Nobel Prize in Physiology in 1904 for his work on digestive secretio. Pavlov provided the most famous example of *classical conditioning*. During his research on the physiology of digestion in dogs, Pavlov noticed that, rather than simply salivating in the presence of food, the dogs began to salivate in the presence of the lab technician who normally fed them. Pavlov called this anticipatory salivation *psychic secretion*. From this observation he predicted that, if a particular stimulus in the dog's surroundings was present when the dog was given food, then this stimulus would become associated with food and cause salivation on its own. In his initial experiment, Pavlov used a bell to call the dogs to their food and, after a few repetitions, the dogs started to salivate in response to the bell. Pavlov called the bell the *conditioned stimulus* (CS) because its effect depended on its association with food. He called the food the *unconditioned stimulus* (US) because its effect did not depend on previous experience. Likewise, the response to the CS was the conditioned response (CR) and the response to the US was the unconditioned response (UR) (Brink, 2008).

The Bell Rings before Food Is Presented

Conditional Response

Pavlovian conditioning, also known as classical conditioning, is a form of learning in which one stimulus, the CS, comes to signal the occurrence of a second stimulus, the US. Classical conditioning has five characteristics: (1) pairing—conditioning is usually done by pairing the two stimuli, e.g., Pavlov presented dogs with a ringing bell followed by food; (2) often forward conditioning—the CS comes before the US, e.g., a bell ring comes before food appears; (3) time-interval—the shorter the time between the CS and US, the quicker the subject learns; (4) repetition—the CS-UC cycles must be repeated over time for the subject to learn; and (5) prediction or association. If the CS is paired with the US as usual, but the US also occurs at other times, so that the US is just as likely to happen in the absence of the CS as it is following the CS, the CS does not "predict" the US. In this case, conditioning fails: the CS does not come to elicit a CR (Rescorla, 1967). This finding regarding the prediction function greatly influenced subsequent conditioning research and theory.

Another property of CR is its reversibility. *Reversibility*, or extinction as it is sometimes called, is the repeated presentation of a CS in the absence of a US (ring a bell but don't provide food). This is done after a CS has been conditioned by one of the methods above. When this is done the CR frequency eventually returns to pretraining levels.

Conditioned drug response, conditioned hunger, and conditioned emotional response are all great examples of classical conditioning. Conditioned response theory has been helpful in studying the sensory abilities of various animals. For example, Karl von Frisch was able to determine that honeybees can see several colors by conditioning them to look for food on blue cardboard. Once they showed the proper conditioned response, he did the same with cardboard in other colors and discovered that bees can tell the difference between blue and green, blue and violet, and yellow and green. An example of CR of humans would be that children first learn to associate the word, "No!" with an angry face, and eventually learn to stop whatever behavior proceeds the "No!" Most conditioned responses, especially those learned at an early age, become permanently ingrained (wisegeek.com). Classical conditioning is widely used in the neural basis of learning and memory and behavioral therapies.

Frederic Skinner followed the idea of conditioned reflexes developed by Ivan Pavlov and applied it to the study of behavior. One of his best known inventions is the Skinner Box. A starved rat was introduced into a box, and when it pressed a lever, a small pellet of food was dropped onto a tray. The rat soon learned that whenever he pressed the lever he would receive some food. In this experiment the lever pressing behavior was reinforced by the presence of food.

If pressing the lever is reinforced (the rat gets food) when a light is on but not when it is off, the rat soon forms discrimination between light and dark, and responses (pressing the lever) continue to be made in the light but not in the dark. Here, Skinner demonstrated the ideas of "operant conditioning" and "shaping behavior." Unlike Pavlov's "classical conditioning," where an existing behavior (salivating for food) is shaped by associating it with a new stimulus (ringing of a bell), operant conditioning is the rewarding of a partial behavior or a random act that approaches the desired behavior (pressing the lever in the light). Operant conditioning can be used to shape behavior. Computer-based self-instruction uses many of the principles of Skinner's technique (www.pbs.org).

7.4   Ants That Count!

A story by Robert Krulwich (2009) tells of an ingenious experiment conducted in the Sahara suggesting that ants may be able to count. Here, as he gave it, is his story (you can see hilarious cartoon movies about this experiment on Youtube.com):

How Do Ants Get Home?
    Most ants get around by leaving smell trails on the forest floor that show other ants how to get home or to food. They squeeze the glands that cover

their bodies; those glands release a scent, and the scents in combination create trails that other ants can follow.

That works in the forest, but it doesn't work in a desert. Deserts are sandy and when the wind blows, smells scatter.

So How Do Desert Ants Find Their Way Home?

It's already known that ants use celestial clues to establish the general direction home, but how do they know the number of steps to take that will lead them right to the entrance of their nest?

Ants cannot find their home with shortened legs.

Ants That Count

Wolf and Whittlinger trained a tribe of ants to walk across a patch of desert to some food. When the ants began eating, the scientists trapped and divided them into three groups. They left the first group alone. With the second group, they used superglue to attach pre-cut pig bristles to each of their six legs, essentially putting them on stilts. The third group had their legs cut off just below the "knees," making each of their six legs shorter.

After the meal and makeover, the ants were released. All of them headed home to the nest while the scientists watched to see what would happen.

The "Pedometer Effect"

The regular ants walked right to the nest and went inside.

The ants on stilts walked right past the nest, stopped and looked around for their home.

The ants on stumps fell short of the nest, stopped, and seemed to be searching for their home.

It turns out that all the ants had walked the same number of steps, but because their gaits had been changed (the stilted ants, like Monty Python creatures, walked with giant steps; the stumpy ants walked in baby steps) they went exactly the distance one would predict if their brains had counted the number of steps to the food and then, after reversing direction, had counted the same number of steps back. In other words, all the ants counted the same number of steps back!

The experiment results suggest strongly that ants have something like pedometers that do something like counting.

Now here is a question for you: How can desert ants find their home if, after they are out for counting the food, wind blows, making a different waved sandy surface such that the path returning to home becomes longer or shorter?

7.5   Disease Outbreak and Network Chaos

According to *ScienceDaily* (Sept., 2010), Nicholas Christakis, a professor of medicine, medical sociology and sociology at Harvard University, and James Fowler, a professor of medical genetics and political science at the University of California, San Diego, used a friendship network of 744 students to study flu epidemic. As the 2009 influenza season approached, they constructed a social network that contacted 319 Harvard undergraduates, who in turn named a total of 425 friends. Monitoring the two groups through self-reporting and data from Harvard University Health Services, the researchers found that, on average, the friends group manifested the flu roughly two weeks in advance of the randomly chosen group, and a full 46 days prior to the epidemic peak in the whole population. The findings are statistically significant (Christakis and Fowler, 2010).

Christakis and Fowler explained the parallel between a general popula-tion and a subgroup in the following way. Just as they come across gossip, trends and good ideas sooner, the people at the center of a social network are exposed to diseases earlier than those at the margins. Traditionally, public health officials often track epidemics by following random samples of people or monitoring people after they get sick. The friendship network has suggested an effective way to find out which parts of the country are going to get the flu first and provides a way we can get ahead of an epidemic of flu, or poten-tially anything else that spreads in networks. Of course, a model is required to generalize results from a small friendship network to the whole population. Now here us a question for you: How will the characteristics of the starting group (diseased or not) affect the outbreak prediction?

A common type of network is the so-called *scale-free network*, whose degree distribution follows a power law, at least asymptotically. That is, the fraction $P(k)$ of those nodes in the network having k direct connections to other nodes is proportional to k^r, where the parameter r typically ranges from 2 to 3. For instance, $\gamma \approx 2.5$ for www, 2.0 for coauthor networks, 2.1 for phone call networks, 2.8 for networks of Words and synonyms, 2.2 for metabolic networks such as those of the bacteria E. coli, and 2.4 for networks of proteins, 3.4 for sexual contacts and the industrially used microorganism S. cerevisiae. A very important property of scale-free networks is their resilience to the deletion of nodes. This means that if a set of random nodes is deleted from a large scale-free network, the network can still function properly because it is composed

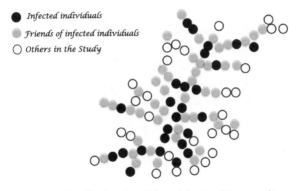

Simplified version of the social network (744 people)
studied by Fowler and Christakis, on Dec. 8, 2009

A Friendship Network: Monitoring a Disease Outbreak

mainly of lower degree nodes, the random deletion of which will have little effect on the overall degree distribution and path lengths. We can see many examples of this on the Internet and the Web. Individual computers on the Internet fail or are removed all the time, but this doesn't have any obvious effect on the operation of the Internet or on its average path length. If one or more hubs are deleted, the network will probably no longer function properly. Mitchell (2009) provides fascinating examples of network failure that made the national news:

> August 2003: A massive power outage hit the Midwestern and Northeastern United States, caused by cascading failure due to a shutdown at one generating plant in Ohio. The reported cause of the shutdown was that electrical lines, overloaded by high demand on a very hot day, sagged too far down and came into contact with overgrown trees, triggering an automatic shutdown of the lines, whose load had to be shifted to other parts of the electrical network, which themselves became overloaded and shut down. This pattern of overloading and subsequent shutdown spread rapidly, eventually resulting in about 50 million customers in the Eastern United States and Canada losing electricity, some for more than three days.
>
> August 2007: The computer system of the U.S. Customs and Border Protection Agency went down for nearly ten hours, resulting in more than 17,000 passengers being stuck in planes sitting on the tarmac at Los Angeles International Airport. The cause turned out to be a malfunction in a single network card on a desktop computer. Its failure quickly caused a cascading failure of other network cards, and within about an hour of the original failure, the entire system shut down.

August–September 1998: Long-Term Capital Management (LTCM), a private financial hedge fund with credit from several large financial firms, lost nearly all of its equity value due to risky investments. The U.S. Federal Reserve feared that this loss would trigger a cascading failure in worldwide financial markets because, in order to cover its debts, LTCM would have to sell off much of its investments, causing prices of stocks and other securities to drop, which would force other companies to sell off their investments, causing a further drop in prices, et cetera. At the end of September 1998, the Federal Reserve acted to prevent such a cascading failure by brokering a bailout of LTCM by its major creditors.

7.6   Technological Innovation

The network is a symbol of our lives today. Effectively using networks to solve societal problems is certainly in everyone's interest. Here are two notable applications.

Link-Analysis: In many situations, finding causal relationships is the goal. When there are a larger number of variables, this task is not trivial. However, association is a necessary condition for a causality relationship. Finding the set of events that correlate many others is often the focus point for further research. Link-analysis explores the associations between large numbers of objects of different types. It can be used to examine the adverse effects of medications in the drug postmarketing surveillance to determine whether a particular adverse event can be linked to just one medication or to a combination of medications. Link analysis can further examine specific patient characteristics to determine whether they might correlate to both the adverse event and to the medication (Cerrito, 2003).

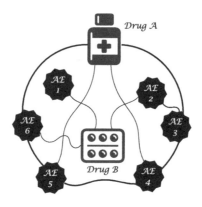

A Link Analysis Can Identify
Drug-AE (Adverse Event) Associations.

Link-analysis provides a way to find the event set with high density. Finding sale items that are highly related or frequently purchased together can be very helpful for stocking shelves, cross-marketing in sales promotions, catalog design, and customer segmentation based on buying patterns. A commonly used algorithm for such density problems is the so-called Apriori (Motoda & Ohara, 2009). As an example of how this algorithm can be applied, suppose it is found that most people who buy peanut butter also like to buy sandwich bread during the same trip to a supermarket. A store manager can use this information in at least two different ways. He can (1) give the customer a convenient and pleasant shopping experience by stocking the two items next to each other, or (2) separate the two products widely, forcing customers to walk across the store, in the hope they will look at, and perhaps buy, more.

PageRank Algorithm: Living in a network, we often want to find significant figures within the network. This is very similar, in a way, to ranking the importance of a webpage. We know that a person who has many connections must be important in some way. We also know that having connections with many "important people" will make one more important than just connecting to "unimportant people." To rank a webpage, we follow the same logic: If a webpage has many visitors, it must be important. If a webpage is hyperlinked by many other webpages, then it will make the original webpage important. If the webpage is hyperlinked by webpages that are hyperlinked by many other webpages, this will make the original webpage even more important. That is the notion of PageRank—a webpage ranking algorithm. The name PageRank is a trademark of Google, who obtained exclusive license rights on the patent from Stanford University (U.S. Patent 6,285,999). The university received 1.8 million shares of Google stock in exchange for use of the patent. PageRank can be computed either iteratively or algebraically. The iterative method can be viewed as a Monte Carlo simulation, which mimics web-visitors surfing over the Internet. The idea of formulating algebraically a link analysis problem as an eigenvalue problem was suggested as early as 1976 by Pinski and Narin, who worked on the ranking of scientific journals.

7.7 Critical Path Analysis

In Chapter 4, we introduced *Bellman's Optimality Principle:* An optimal policy has the property that, whatever the initial state and initial decision are, the remaining decisions must constitute an optimal policy with regard to the state resulting from the first decision. We now study its application in a decision network, the so-called critical path analysis (CPA). CPA is a commonly used tool in project management, in which sequentially dependent tasks are involved and the minimum time to finish the project is of great interest.

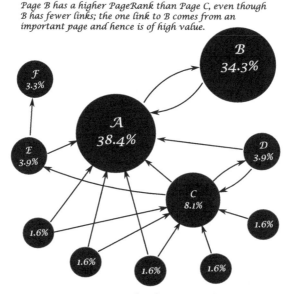

Page B has a higher PageRank than Page C, even though B has fewer links; the one link to B comes from an important page and hence is of high value.

An Example of PageRanks

Suppose we want to find the shortest path among several alternatives in a network with vertices $1, 2, ..., n$, where each vertex represents an event, e.g., arrival at a bus stop. Certain pairs of vertices (i, j) are directly linked by an arc. Such an arc represents an activity with an associated positive value (length) d_{ij}. A path is a sequence of arcs and its length is the sum of the corresponding d_{ij}. The task is to find the shortest path from vertex 1 to vertex n. Let f_i be the length of the shortest path from vertex i to vertex n. Operationally, Bellman's Optimality Principle can be stated thusly: the length of the shortest path (f_i) starting from vertex i is the smallest sum of the distance from vertex i to vertex j and the shortest path (f_j) from vertex j among all vertices j which are directly linked to vertex i. Mathematically it can be stated as: $f_n = 0$ and for each $i < n$,

$f_i = $ minimum of $d_{ij} + f_j$ among all vertices j *directly linked* to vertex i.

It is easy to develop a backward induction algorithm for finding shortest path based on this formulation. The solution, giving the length of the shortest path to vertex 10 for each i, is found by working from right to left on the network diagrammed below (You should mark f_i one by one on the figure at each vertex to help you understand. Remember, f_i is the shortest distance from vertex i to vertex 10.):

$i =$	10	9	8	7	5	4	3	2	6	1
$f_i =$	0	5	9	8	13	12	15	15	17	17

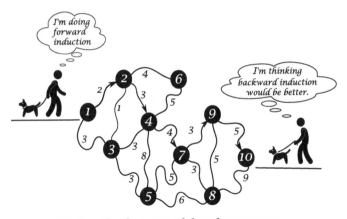

Finding the Shortest Path from Vertex 1 to 10

The solution indicates a direction as shown by the arrows. Thus, the shortest path from 1 to 10 is: $1 \to 2 \to 4 \to 7 \to 9 \to 10$, and its length is 17.

The events (vertices) on the critical path are often called milestones. They are critical to the success of the entire project. After the critical path and the starting time of the first event is determined, schedules for the milestones can be directly calculated. Activities that are not on the critical path or paths might be slightly delayed or rescheduled as a manager desires without delaying the entire project.

Critical path analysis is treated as an application of *linear programming* in Operation Research, rather than sequentially or by backward induction as discussed above.

7.8   Revelations of the Braess Paradox

We are going to discuss three examples of enlightenment resulting from the consideration of Braess's Paradox.

Imagine a publishing company that has two types of printers and book-binding machines: small and large. The cost is \$10 per book for the large printer. For the small machine, the cost increases as the daily volume n increases because it loses efficiency in handling larger volumes. The cost is \$2 + 0.07n$. Similarly, the cost is \$10 per book for the larger binding machine and \$2 + 0.07n$ for the small binding machine. There are 100 books of various kinds to be printed and bound daily. Departments handling different kinds of books have been asked to reduce their cost as much as possible as cost reductions will be tied to their annual bonuses.

After simple calculations all the department heads decided to use the small printer and small binding machine. Their reasoning notes that the cost for a large printer or binding machine is $10 per book, whereas in the worst case scenario the cost for the small printer or binding machine is $2 + 0.07(100) = $9 per book. As such, the cost for each is at most $18 per book. One hundred books will cost the company $1800 daily. However, we have learned from the Braess Paradox that we can do better than this. We can suggest the company implement two production lines: (1) a small printer and large binding machine and (2) a large printer and small binding machine. With the option of these two production lines, when each department acts rationally (i.e., to minimize its costs), the system eventually reaches the Nash equilibrium with 50% books in each production line and the cost will be $10 + 2 + 0.07(50) = $15.50 per book. One hundred books will cost the company $1550 daily, a $250 or 14% savings from the original budget of $1800. There is also a time-savings in comparison to the case when only small apparatuses are used.

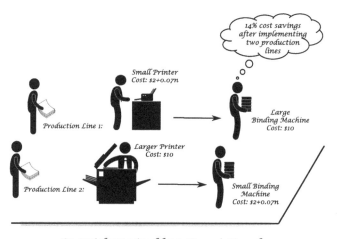

An Insight Derived from Braess's Paradox

In many large cities motorists complain of horrible traffic conditions, and worry that they will only get worse. Advertisements heartily recommend that people use public transportation more, but few of them actually do so; drivers feel it is relatively convenient to drive instead of taking public transportation. Let's presume that taking public transportation for a specific route results in a travel time of 49 minutes, and that when n people drive cars, the travel time is $25 + 0.001n$ minutes. Suppose there are $20,000$ car owners who either drive alone in their cars or take buses as daily transportation. If everyone drives a car the travel time will be 45 minutes on average, less than the 49 minutes by bus. Therefore, everyone prefers to drive.

Now suppose that, after some analysis, the city government takes an action: Cars with even plate numbers can be driven only on even calendar dates (e.g., March 2, 4, 6, ...), and cars with odd plate numbers only on odd calendar days (March 1, 3, 5, ...). This will lead to 10,000 cars on the road and 10,000 bus riders each day with an average travel time of $(35 + 49)/2 = 42$ minutes. The daily 3-minute savings in travel time per person also protects the environment. Of course, the reality is that there are many other complications in the implementation of such a policy.

Farmers always need to decide what to grow or raise to maximize their profits. Let's say, for example, the net profit for raising cows is $80\% - 30\%p$ in five years, where p is the proportion of farmers in the cow business. The net profit for another kind of farming (assume that the farmer has just one option besides raising cows) is $48\% - 5\%(1 - p)$ in five years. Here $1 - p$ is the proportion of farmers choosing the alternate livelihood. Therefore, the net profit for a cow farmer is at least 50% in five years, while the profit for the other business is at most 48%. Every farmer will naturally choose to raise cows! However, this is not optimal. Suppose, as an example, the government provides a five-year incentive program for entering into the alternate business, inducing farmers to switch every five years from cow farming, so that about $p = 60\%$ farmers will be in cow business and the average profit they can generate is $(0.8 - 0.3p) p + (0.48 - 0.05(1 - p)) (1 - p) = 55.6\%$. To have this maximum profit, cooperation among farmers is necessary to keep the right proportion of people in the business of raising cows.

7.9 Artificial Swarm Intelligence

A *social animal* or *social insect* is an organism that is highly interactive with other members of its species to the point of having a recognizable and distinct society, in which interactions and cooperation go beyond familiar members.

Swarm Intelligence can be found in social insects. Colonies of ants organize shortest paths to and from their foraging sites by leaving pheromone trails, by forming chains of their own bodies to create bridges and to pull and hold leaves together with silk, and by dividing work between major and minor ants. Similarly, bee colonies regulate hive temperature and work efficiently through specialization (division of labor).

Social insects organize themselves without a leader. The modeling of such insects by means of self-organization can help design artificial swarm-intelligent systems. For instance, ant colony optimization and clustering models are inspired by observed food foraging and cleaning activities in ant colonies; a particle swarm optimization model imitates the social behavior of fish schools and bird flocks; and the bird flocking model for documentation clustering was inspired by bird flock flying behaviors. In a natural swarm intelligence system direct interaction and communication occurs through food exchange, visual contact, and chemical contact (pheromones). Such interactions can be easily imitated in artificial swarm-intelligence. In addition to direct interaction there can be indirect interactions among members where individual behavior modifies the environment, which in turn modifies the behavior of other individuals (stigmergy).

(1) Ant Routing Algorithm for Optimization

An ant routing algorithm, introduced by Marco Dorigo (1992), was inspired by the food foraging behavior of ants hunting the shortest or fastest route. Its key algorithm can be described thusly:

(1) Ants lay pheromones on the trail when they move food back to nest.
(2) Pheromones accumulate with multiple ants using the same path, evaporating when no ants pass by.
(3) Each ant always tries to choose trails having higher pheromone concentrations.
(4) In a fixed time period, ant agents are launched into a network, each agent going from a source to a destination node.
(5) The ant agent maintains a list of visited nodes and the time elapsed in getting there. When an ant agent arrives at its destination, it will return to the source following the same path by which it arrived, updating the digital pheromone value on the links that it passes by. The slower the link, the lower the pheromone value.
(6) At each node, the ant colony (data package) will use the digital pheromone value as the transitional probability for deciding the ant (data) transit route.

Ant routing algorithms can be used to better route traffic on telecommunications systems and the internet, roads, and railways to reduce congestion that efficient routing algorithms may provide. Southwest Airlines is actually putting the ant colony research to work, with an impressive payback.

(2) Ant Clustering Model for Data Clustering

The size of the internet has been doubling in size every year. Organizing and categorizing the massive number of documents is critical for search engines. The operation of grouping similar documents into classes can also be used to obtain an analysis of their content. Data clustering is the task that seeks to identify groups of similar objects based on the value of their attributes. The ant clustering algorithm categorizes web documents into different domains of interest.

Ant Clustering Model in Action

In nature, ants collect and pile dead corpses to form "cemeteries." The corresponding agent (ant) action rules are: agents move randomly, only recognizing objects immediately in front of them, picking up or dropping an item based on the pickup probability $P_p = \left(\frac{c_p}{c_p + f}\right)$ and drop off probability $P_d = \left(\frac{c_d}{c_d + f}\right)$, where f is the fraction of the items in the neighborhood and c_p and c_d are constants.

In PSO (Particle Swarm Optimization), individuals are the "particles" and the population is the "swarm." PSO is based on simple social interactions, but can solve high-dimensional, multimodal problems. In PSO systems, every particle is both teacher and learner. The communication structure determines how solutions propagate through the population. After a solution has been learned by the population, resilience allows the particles to start exploring again, should they receive substantially new information. Resilience versus efficiency is often a trade-off.

(3) The Flocking Model

The flocking model, one of the first bio-inspired computational collective behavior models, was first proposed by Craig Reynolds in 1987. There is no central control in a bird flock. Each bird acts based on three simple behaviors: (1) the bird alignment—steers towards the average heading of its local flock mates, (2) separation—steers to avoid crowding flock mates, and (3) cohesion—steers towards the average position of its local neighbors.

In the flocking model a problem with a quantifiable solution is found by considering the individual solutions in time of each individual in a flock. A particle (individual) is composed of three vectors (three sets of data): The x-vector records the current position (location) of the particle in the search space, the p-vector records the location of the best solution found so far by the particle, and the v-vector contains a gradient (direction) in which the particle will travel if undisturbed.

For population evolution, there are also two fitness values: the x-fitness records the fitness of the x-vector and the p-fitness records the fitness of the p-vector. A particle's status can be updated synchronously and asynchronously. Asynchronous updates allow newly discovered solutions to be used more quickly. James Kennedy and Russ Eberhart (1995) further develop the PSO model that incorporates a social component.

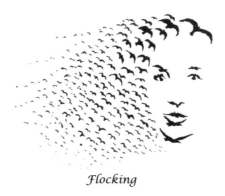

Flocking

Swarm intelligence-based applications include complex interactive virtual environment generation in the movie industry, cargo arrangement in airline companies, route scheduling for delivery companies, routing packets in telecommunication networks, power grid optimization controls, data clustering and data routing in sensor networks, unmanned vehicle control in the U.S. military, and planetary mapping and microsatellite control as used by NASA (Xiaohui Cui, ppt).

7.10   One Stone, Three Birds

We are going to give several examples of how the same methods can be used for problems that appear to be completely different.

Mark–Recapture Methods (MRM) have a long history, having been created for the study of fish and wildlife populations before being adapted for other purposes. The application of these methods to the study of epidemiologic problems came relatively late in this history and thus has been able to draw on advances in the other areas as well as in statistical methods more broadly. The simplest capture–recapture model is the so-called two-sample model, used solely to estimate the unknown size of a population. The first sample provides individuals for marking or tagging and is returned to the population, while a later, second sample consists of the recaptures. Using the numbers of individuals caught in both samples, we can estimate the total population size.

Suppose we want to estimate the number of fish in a pond without pumping out the water. We capture m fish from the pond, mark them, and put them back in the pond. On the next day, we catch some fish from the same pond, from which we find that some proportion, p, are marked. Because p is expected to be $\frac{m}{n}$, the total n can be estimated by $n = \frac{m}{p}$.

This is sometimes called the Lincoln–Petersen method. It can be used to estimate population size if (1) only two independent samples are taken, (2) each individual has an equal probability of being selected (or captured), and (3) the study population is closed, meaning that no deaths, births, or migration occur in the interim between the two samplings.

MRM is a method commonly used in ecology to estimate population size and vital rates, i.e., survival, movement, and growth. In medicine it can be applied to estimate the number of people needing particular services (e.g., people infected with HIV). MRM can also be used to study the error rate in software source code, in clinical trials, databases, etc. This approach is useful when the total population size is unknown.

Suppose now we want to measure the volume of water, V, in a pond. Making an analogy to the problem of estimation of the number of fish in the pond, we can use MRM: We pour a cup, of volume v, of a colored, nontoxic testing liquid into the pond and let it sufficiently mix with the pond water through diffusion. We then take a cup of water from the pond and measure the concentration, c, of the test liquid. Because c is expected to be $\frac{v}{V}$, the volume of the water in the pond can be estimated as $V = \frac{v}{c}$ (we have assumed v is very small and negligible in comparison with V).

Let's look at another example of how to use one method to solve multiple, interrelated problems: the calculation of a country's territory, the evaluation of constant π, and an integration problem in calculus.

Suppose we are asked to find the size of the USA's territory. The only "data" we have is a map, in 1:20000 scale, on a unit square piece of white paper. We can accomplish our mission as follows.

Assume the USA territory in the map has an area S and is colored. We tear the map into small pieces, mix them up, and randomly choose some. Suppose we find that a proportion p of the pieces are colored. Since the paper is a unit square, p is expected to be S, so we estimate the area S by p. Because the map is in a 1 : 20000 scale, the actual size of USA territory is therefore $20000^2 p$.

But what if we don't what to destroy the map? Well, that is easy too. We find a small dart-like object, perhaps, a needle, which we can throw randomly at the map many times, recording the proportion p of times when the needle tip lands in the USA map among the total times when the needle lands

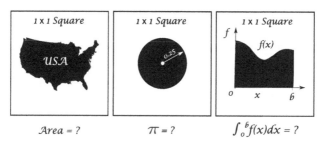

Three Birds with One Stone

somewhere in the unit square of paper. We can even use computer simulation to determine p or the area S.

We now modify the problem slightly by replacing the map with a disk of radius 0.25. Then the area of the disk is $0.25^2\pi = p$, from which we find that the constant $\pi = 16p$. In other words, we can use (computer) simulation to determine the constant π. Furthermore, since a definite integral of a positive function $f(x)$ is the area under the curve $f(x)$, $\int_\Omega f(x)\,dx = p$. Therefore, a simple (computer) simulation can be used to evaluate the integral of a function that represents the boundary of a region.

7.11 Scaling in Biology

Biological Scaling is the study of how the biological properties of an organism change as its size or mass changes. Since metabolism, the conversion by cells of food, water, air, and light to usable energy, is the key process underlying all living systems, this relation is enormously important for understanding how life works (Mitchell, 2009). How the basal metabolic rate (the average energy used by an organism while resting) scales with the organism's body mass is an interesting and fundamental topic of biological scaling. In 1883, German physiologist Max Rubner proposed a scaling law based on the second law of thermodynamics. Max reasoned that an organism's metabolic rate can be defined as the rate at which its cells convert nutrients into the energy used for all the cell's functions and for building new cells. In the process, the organism gives off heat as a by-product. An organism's metabolic rate can thus be inferred by measuring this heat production. Furthermore, to keep the body at a constant temperature, the heat produced by the metabolism must equal the heat loss through its external (skin) and internal (lungs) surface areas. The rate of such heat loss is proportional to the surface area, whereas the rate produced is proportional to body mass (M). Body mass is proportional to L^3 and surface area is proportional to L^2, where L is the characteristic length of the organism, assuming the organisms are geometrically similar. Therefore, to maintain the body temperature, the rate of heat generation (the metabolic rate) should to proportional to the body surface or L^2 or $M^{2/3}$.

Rubner's theory was accepted for next 50 years, until the 1930s, when a Swiss animal scientist, Max Kleiber, set out to conduct experiments measuring metabolism rates of different animals. Kleiber's results showed that metabolic rate scales with body mass to the three-fourths power: that is, metabolic rate is proportional to $M^{3/4}$. This leads to the so-called *metabolic scaling theory*. Since metabolic rate is the rate at which the body's cells turn fuel into energy, biologists (West, Brown, and Enquist, 1999) set out to justify Kleiber's law using fractal geometry. Their main assumption was that evolution has produced circulatory and other fuel-transport networks that are maximally "space filling" in the body—that is, that can transport fuel to cells in every part of the body.

Again, the quarter-power scaling laws are not as universal as the theory claims because animals are not geometrically similar, neither their habits nor their behaviors.

We now study a slightly more complicated case of scaling using dimensional analysis (Section 2.11) to help predict the most comfortable walking speeds of human beings. The example is adapted from the example by Szirtes (2007, pp. 518–523)

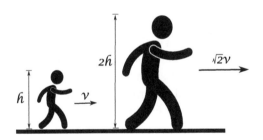

The Most Comfortable Speed to Walk

It is easy to observe that the most comfortable walking speeds of human beings depend on their size. In general, adults walk faster than children (when they walk alone). An interesting question is: How does size affect the walking speeds of geometrically similar human beings? Although one can prefer to walk slower or faster, there is a particular speed at which the energy expended per unit distance traveled is minimum, the most comfortable walking speed. To investigate this, we can use dimensional analysis.

Let the most comfortable walking speed v be the dependent variable with meters per second, $m \cdot s^{-1}$, the dimension of characteristic linear size (e.g., height) L be m, and the dimension of gravitational acceleration g be $m \cdot s^{-2}$. We included gravitational acceleration because in walking the center of mass of the body also moves vertically, and so gravitation affects the expended energy. The inclusion of gravitational acceleration also allows the results to generalize to different celestial bodies (e.g., the Earth or the Moon). There are three variables and two dimensions; hence, we have only one dimensionless variable $\pi_1 = \frac{v}{\sqrt{Lg}}$, which must be some constant c. Therefore, we obtain $v = c\sqrt{Lg}$. This implies that the most comfortable walking speed is proportional to the square root of a person's characteristic length, or height. Now in addition to geometric similarity we assume, as a first approximation, that human beings are equally dense such that the body mass M is proportional to L^3. This leads to the conclusion that v is proportional to $M^{1/6}$. Thus, the most comfortable walking speed is proportional to the sixth root of the individual's mass.

One can further investigate oxygen consumption and other variables, we do not do so here.

7.12   Genetic Programming

Genetic programming (GP), inspired by our understanding of biological evolution, is an evolutionary computation (EC) technique that automatically solves problems without requiring the user to know or specify the form or structure of the solution in advance. At the most abstract level GP is a systematic, domain-independent method for getting computers to solve problems automatically starting from a high-level statement of what needs to be done (Poli, Langdon, and McPhee, 2008). The idea of genetic programming is to evolve a population of computer programs. The aim and hope is that, generation by generation, GP techniques will stochastically transform populations of programs into new populations of programs that will effectively solve problems under consideration. Similar evolution in nature, GP has in fact been very successful at developing novel and unexpected ways of solving problems.

(1) Representation: *Syntax Tree*

To study GP, it is convenient to express programs using syntax trees rather than as lines of code. For example, the programs $(x + y) + 3$, $(y + 1) \times (x/2)$, and $(x/2) + 3$ can be represented by the three syntax trees in the diagram, respectively. The variables and constants in the program (x, y, 1, 2, and 3) are leaves of the tree, called *terminals*, while the arithmetic operations ($+$, $\times$ and *max*) are internal nodes, or *functions*. The sets of allowed functions and terminals together form the primitive set of a GP system.

(2) Reproductive Mechanism

For programs to propagate, GP must have well-defined mechanisms of reproduction. Here, there are two common ways to generate offspring: crossover and mutation. *Crossover* is the primary way to reduce the chance of chaos because the crossovers lead to much similarity between parent and child.

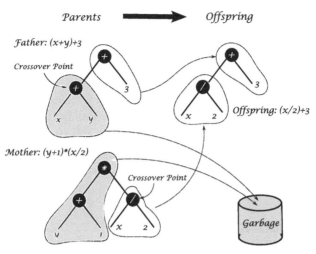

Crossover in Genetic Programming

Crossover involves a selection of a subtree for crossover, whereas mutation randomly generates a subtree to replace a randomly selected subtree from a randomly selected individual. Crossover and mutation are illustrated in the diagram above, where the trees on the left are actually copies of the parents; their genetic material can freely be used without altering the original individuals.

(3) Survival Fitness

The second mechanism required for program evolution is *survival fitness*. There are many possible ways to define a measure of fitness. As an example, for the problem of finding a function $g(x)$ to approximate the target function $f(x)$, the mean square error between the two functions can be used as the fitness measure.

The usual termination criterion is that an individual's fitness should exceed a target value, but it could instead be a problem-specific success predicate or some other criterion. Typically, a single best-so-far individual is then harvested and designated as the result of the run.

Koza (Banzhaf, et al., 1998) studied the symbolic regression

$$y = f(x) = \frac{x^2}{2}, \text{ where } x \text{ ranges from 0 to 1.}$$

Using as the terminal set x ranging from -5 to 5 on the syntax tree; a function set consisting of $+, -, \times$, and protected division %; and 500 individuals in each generation, Koza was able to obtain the best individual (function) f_i in generation i, where $f_0 = \frac{x}{3}$, $f_1 = \frac{x}{6-3x}$, ..., and $f_3 = \frac{x^2}{2}$. Therefore, at generation 3, the correct solution is found. However, as the generation number increases, the best fit function starts to expand again.

Genetic programming does not have to be only for mathematical function fitting. In fact it has proliferated with applications in many fields, including: RNA structure prediction (van Batenburg, Gultyaev, and Pleij, 1995), molecular structure optimization (Willett, 1995; Wong et al. 2011; Wong, Leung, and Wong, 2010), code-breaking, finding hardware bugs (Iba et al., 1992), robotics (Li et al. 1996; Michail, 2009), mobile communications infrastructure optimization, mechanical engineering, work scheduling, the design of water distribution systems, natural language processing (NLP), the construction in forensic science of facial composites of suspects by eyewitnesses (Li et al., 2004), airlines revenue engagement, trading systems in the financial sector, and the audio industry.

7.13   Mechanical Analogy

Analogies play a central role in scientific research. In section 7.10, we showed an example of how to "kill three birds with one stone." Here we give examples to show how to use instruments to mechanically solve optimization problems.

In Section 7.7 we discussed the critical path analysis to find the shortest path in a network from the starting point 1 to the destination point 10. If we can physically make a net with strings and each segment of the strings has the length equal to the distance specified between the two knots, we can hold the two knots 1 and 10, and pull from two opposite directions until the net is as tight as possible. The path along all the tightest strings from knots 1 to 10 is the shortest path we are seeking. (Can you see why?)

Apparatuses can be used to solve other optimization problems. Suppose a city decides it wants to build a new library. The only criterion of selecting its location is to minimize the travel time of the people living in the four areas of the city. To solve this problem mechanically, we can mark the central locations of the four areas on a piece of cardboard, or on a wall, and hammer nails or install pulley holes at the four locations; use strings to tie four different weights according to the population size of each area; pass the strings through the four different pulleys, separately, and tie the other ends of the four strings together. When you loosen your grip, the knot starts to move and quickly it will stop at a location, the location for the library which minimizes the total travelling time for all the people in the town. (Why?).

Apparatuses can be used to solve many other problems. Here's an empirical method of approximating the geometrical center of a convex object that is nearly two-dimensional (lying almost in a plane with a tiny and nearly constant thickness), assuming also that its mass is evenly distributed. First, hold the object by a string attached to any point on its edge; mark a line along the string on the object when it is balanced. Pick another point near the edge

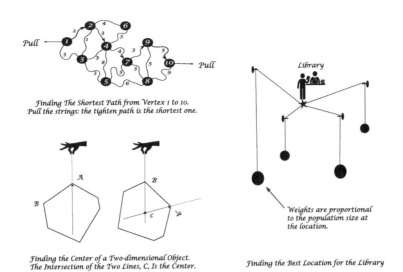

Finding The Shortest Path from Vertex 1 to 10.
Pull the strings: the tighten path is the shortest one.

Finding the Center of a Two-dimensional Object.
The Intersection of the Two Lines, C, Is the Center.

Weights are proportional
to the population size at
the location.

Finding the Best Location for the Library

Mechanism of Physical Instruments in Solving Problems

and do the same thing again. So we have two lines drawn on the object. The intersection of the two lines is the center of gravity of the object. This can be proved mathematically since the center is the location about which the first moment of the mass of the object is minimized, and the center of gravity and geometrical center are identical for a homogeneous object of any shape.

7.14   Numerical Methods

The *finite element method* (FEM) in applied mathematics is among the most commonly used techniques when making complex engineering and scientific calculations. In fact, many great achievements in modern engineering and science would be impossible without such methods. FEM is a numerical technique for finding approximate solutions to differential equations, which are equations involving the (partial) derivatives of unknown quantities. *Differential equations* often arise when researchers attempt to solve complex physical problems in the aerospace and automotive industries, in civil engineering and weather prediction, and elsewhere. FEM allows us to break down such a problem into many simpler interrelated problems. This can be achieved by dividing a domain of the original differential equation into many subdomains, called finite elements, each of which can be modeled by an equation much simpler than the original one. The simpler equations can then be combined into a system of linear equations by incorporating the continuity and other mechanical conditions existing at the interfaces between elements.

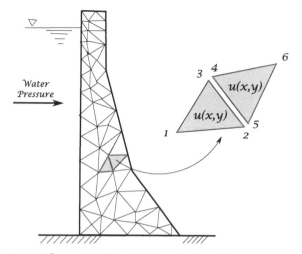

Finite Element Meshes of a Cross-Section of an Arc Dam

A primary idea behind the finite element method is to use an aggregation of simple piecewise local functions to approximate the global complex function, just as we use many small plates to form a sphere, as we see in the way a soccer ball is made. Typically, two steps are involved in FEM: (1) construct a finite element mesh by dividing the domain of the problem into a collection of subdomains or finite elements; over each element, a simpler equation is used to approximate the original equation; and (2) assemble the equations from all elements into a global system of linear equations.

To illustrate the concept of FEM, we number the nodes of a typical tri-angular finite elements as 1, 2, and 3. The unknown function $u(x, y)$, defined on this element, can be approximated by a linear function of the coordinates (x, y) of the element, i.e., $u(x, y) = ax + by + c$, where the constants a, b, and c are dependent on the values of the function at the three nodes, i.e., $u(x_1, y_1)$, $u(x_2, y_2)$, and $u(x_3, y_3)$. The three constants can be easily solved by replacing the three node-values into the equation $u(x, y) = ax + by + c$. But $u(x_1, y_1)$, $u(x_2, y_2)$, and $u(x_3, y_3)$ are unknown yet unless the nodes are on the boundary of the original domain. To solve the values at the inner nodes we have to assemble all the u values at all elements. Let's look at an adja-cent element with nodes 4, 5, and 6 as shown in the figure above. For this adjacent element we again let $u(x, y) = ax + by + c$; however, this time the coefficients a, b, and c are dependent on $u(x_4, y_4)$, $u(x_5, y_5)$, and $u(x_6, y_6)$. To assemble the system of linear equations, we use the adjacent conditions: $u(x_2, y_2) = u(x_5, y_5)$, $u(x_3, y_3) = u(x_4, y_4)$, and the forces at node 2 and 5 are equal but act in opposite directions; similar conditions hold between nodes 3 and 4. Furthermore, $u(x_6, y_6)$ is a known value at boundary (for a so-called first type boundary value problem, all boundary values of the unknown function u are given). If we use up the adjacency properties between all adjacent elements, the values of $u(x,y)$ at inner nodes can be expressed in terms of all boundary values of $u(x,y)$. Thus, we eventually turn the original problem into a large linear system $Au = v$, where A is a squared matrix whose elements are dependent on the mesh of the domain, the vector v consists of the given values of u at the boundary nodes, and the vector u consists of the unknown values of $u(x,y)$ at inner nodes. The system of linear equations can solved using a PC without any difficulty.

For large complex objects, a larger number of elements is required; especially at the place where the $u(x,y)$ are expected to change sharply, many small elements are necessary. The *boundary element method* (BEM) is a further development from FEM. Since most differential equation can be converted into integral equations where only the boundary values of the unknown function and its derivatives are involved, no inner meshes are needed; one only needs the boundary to be divided into elements (one-dimensional line elements instead two-dimensional finite elements in this case) as the name indicates. In each element a simple function form is assumed. The system of linear equations can be formulated in a similar way as in FEM. Since the number of elements are much fewer in BEM than in FEM, BEM is often more efficient when the domain is homogeneous (of uniform material).

A closely related technique is the random walk method, which can be viewed as a "reverse engineering" of the Laplace equation governing Brownian motion. We know that temperature is determined by the random motion of molecules: the faster the random motion, the higher the temperature will be. The Brownian motion equation is a mathematical characterization of the random motion at the microscopic scale, while at the macroscopic scale the average movement of continuous media (in contrast to the discrete individual molecules) can be characterized by Laplace's partial differential equation. Suppose we want to know the temperature distribution in the two-dimensional object, where the temperature on the boundary is given and there is no heat source inside the object. Problems like this can be modeled by a well-known partial differential equation first considered by Laplace (the proponent of Bayesian techniques mentioned earlier in this chapter.) Instead of using FEM or BEM to solve the *Laplace equation* numerically, we can solve the problem with the assistance of a rather inebriated person. Here is how. Let $u(A)$ denote the temperature at any point A. We ask a drunken man to walk in a random direction with a step length constrained by the circle tangent to the boundary of the domain (see the figure). He continues the random walk $(A \rightarrow B \rightarrow C \rightarrow D_i)$ until he reaches or very nearly reaches a boundary point D_i. We then lead him back to point A, where he starts to randomly walk again, eventually reaching another boundary point. After he

The temperature at the point A is the average temperature at n boundary points reached by the drunken man through his n random walks.

On the boundary the temperature is known.

Temperature of the object varies from point to point: $u(A)$

Random Walk Method in Action

has tried the random walk n times (with our help), the temperature $u(A)$ is in fact equal to the average temperature on the n boundary points he has reached:

$$u(A) = \frac{u(D_1) + u(D_2) + ... + u(D_n)}{n}.$$

Of course we would not let a drunken man actually walk for us; instead we only use a computer to generate random numbers to mimic his walk. Such computer simulation methods are called Monte Carlo simulations (Chang 2010). They have a long history and are increasingly popular in engineering, science and mathematics, mainly due to the development of high speed computers. Laplace's equation can be solved effectively using other methods. We used this example to illustrate the interesting idea of the basic random walk method.

7.15   Pyramid and Ponzi Schemes

Scientific thinking can be used in daily life too, at least by sober individuals. Here is a simple example I encountered 23 years ago.

A *pyramid scheme* is an unsustainable business model that involves promising participants payment or services, primarily for enrolling other people into the scheme, rather than supplying real investments or the sale of products or services to the public. To enhance credibility, most such scams are well equipped with fake referrals, testimonials, and information. Various forms of pyramid schemes are illegal in many countries.

I don't remember how I got a letter from an associate professor of Sociology shortly after I arrived at the school in U.S., some twenty years ago, telling me that there was a good way to make my life the way I wanted. I might have believed him simply because he was a professor, even though I didn't know exactly what I wanted for my life. He came to my apartment from out of state and brought tapes and books, showing me and my wife how easy it would be to make money through Amway, a pyramid scheme. I don't remember the details of what he presented that day, but the key mechanism was as follows. First, I was to pay a membership fee. For that I would get discount prices for things I could purchase from the Amway catalog. Furthermore, if I recruited friends into Amway I would get a cut from the sales they made and, even more attractive, a smaller cut from the sales that are made by the friends of my friends, and so on. He showed an example of a typical salesman who recruits 6 friends. Recruiting 6 friends seems not a very challenging thing to do. He continued: each of my six friends also recruits 6 of his friends, and so on. This sounds truly like a brilliant idea: what I need to do is successfully recruit 6 people and then sit back and do nothing but start to collect money and get rich. Indeed, he showed me a nice catalog describing people who made themselves very rich through Amway, including their lifestyles and testimonies.

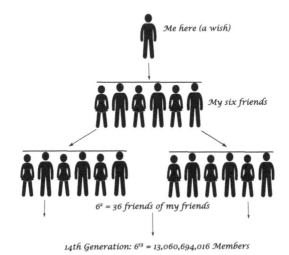

Me here (a wish)

My six friends

$6^2 = 36$ friends of my friends

14th Generation: $6^{13} = 13,060,694,016$ Members

After 14 Generations: All people in the World are Amway Members

I did not do anything about Amway after the professor left and a year later I returned his Amway training materials. The reason I didn't want to get involved with the "business" was that I was busy with school work, but more importantly, the business model bothered me logically and mathematically. Logically, if it is so easy to get rich, where does the money come from? If there are winners, there must be losers too because there is no value creation or savings here! Mathematically, if everyone recruits six different friends, then after 14 "generations" Amway will have recruited everyone in the world: nearly 16 billion people, larger than the total world population of 7.2 billion in 2013. Therefore, recruiting 6 people is not as easy as it appears to be, and only the top people can make it rich. I am sure there are many different ways people have used to figure out the trap in the pyramid scheme. Do I feel lucky for not having been involved in this kind of enterprise? Yes, I am lucky for not being involved in such an unfaithful business, but I would also say: The way I think can only and at most lead me to become a good scientist, but can never make me a successful businessman.

In 2003, the United States Federal Trade Commission (FTC) disclosed what it called an Internet-based "pyramid scam." Its complaint states that customers would pay a registration fee to join a program that called itself an "internet mall" and purchase a package of goods and services such as internet mail, and that the company offered "significant commissions" to consumers who purchased and resold the package. The FTC alleged that the company's program was in fact a pyramid scheme that did not disclose that most consumers' money would be kept, and that it gave affiliates material that allowed them to scam others. Many Facebook games operate on systems similar to pyramid schemes where the player is greatly hindered in the game unless he recruits friends and family to play the game too.

Multi-level marketing involves gains that combine the profits from direct sales of products or services and recruiting salespersons. However, when the profit from the recruitment becomes the majority of profit in the business, it is in fact a pyramid scheme.

A *Ponzi scheme*, named after Charles Ponzi, who became notorious for using the technique in 1920, is a fraudulent investment operation that pays returns to its investors from their own money or the money paid by subsequent investors, rather than from profit earned by the individual or organization running the operation. The Ponzi scheme usually entices new investors by offering higher returns than other investments, in the form of short-term returns that are either abnormally high or unusually consistent. Perpetuation of the high returns requires an ever-increasing flow of money from new investors to keep the scheme going.

When a Ponzi scheme is not stopped by the authorities, it sooner or later falls apart when a large portion of investors take all or a large portion of the remaining investment money. The Madoff investment scandal during the market downturn of 2008 is a well-known Ponzi scheme.

Bernard Lawrence "Bernie" Madoff is an American convicted of fraud and a former stockbroker, investment advisor, and financier. He is the former non-executive Chairman of the NASDAQ stock market, and the admitted operator of a Ponzi scheme that is considered to be the largest financial fraud in U.S. history.

As described at wikipedia.org, Madoff founded the Wall Street firm Bernard L. Madoff Investment Securities LLC in 1960, and was its chairman until his arrest on December 11, 2008. The firm was one of the top market maker businesses on Wall Street, which bypassed "specialist" firms by directly executing orders over the counter from retail brokers.

On December 10, 2008, Madoff's sons told authorities that their father had confessed to them that the asset management unit of his firm was a massive Ponzi scheme, and quoted him as describing it as "one big lie". The following day, FBI agents arrested Madoff and charged him with one count of securities fraud. The U.S. Securities and Exchange Commission (SEC) had previously conducted investigations into Madoff's business practices, but had not uncovered the massive fraud.

On March 12, 2009; Madoff pleaded guilty to 11 federal felonies and admitted to turning his wealth management business into a massive Ponzi scheme that defrauded thousands of investors of their money. The amount missing from client accounts, including fabricated gains, was almost $65 billion. The court-appointed trustee estimated actual losses to investors of $18 billion. On June 29, 2009, Madoff was sentenced to 150 years in prison, the maximum allowed.

Madoff admitted during his March 2009 guilty plea that the essence of his scheme was to deposit client money into a Chase account, rather than invest it and generate steady returns as clients had believed. When clients wanted their money, "I used the money in the Chase Manhattan bank account that belonged to them or other clients to pay the requested funds," he told the court.

7.16 Material Dating in Archaeology

Many disciplines within archaeological science are involved in the proper dating of evidence. The dating of material drawn from archaeological records is done mainly post-excavation, but, to support good practice, some preliminary work called "spot dating" is usually run in tandem with excavation. Dating is obviously very important to the archaeologist who, working to construct an accurate model of the past, must rely on the integrity of datable objects and samples.

There exist many dating methods appropriate to different kinds of materials. Among the main types one finds radiocarbon dating for organic materials, *thermoluminescence dating* for inorganic material including ceramics, *optical dating* for archaeological applications, *potassium–argon dating* for the dating of fossilized hominid remains, and *rehydroxylation*, another method for dating ceramic materials. *Dendrochronology*, or tree-ring dating as it is sometimes called, is the science of dating trees and objects made from wood. *Archaeomagnetic dating* is another interesting method: clay-lined fire hearths take on a magnetic moment pointing to the North Pole each time they are fired and then cool. The position of the North Pole for the last time the fire hearth was used can be determined and compared to charts of known locations and dates.

Radiocarbon dating (or simply carbon dating) is a radiometric technique that uses the decay of carbon-14 ($_{14}C$) to estimate the age of organic materials up to about 60,000 years. The Earth's atmosphere contains the stable *isotope carbon-12* ($_{12}C$) and an unstable isotope carbon-14, roughly in constant proportions. Through photosynthesis, plants absorb both forms from carbon dioxide in the atmosphere. When an organism dies, it contains the standard ratio of $_{14}C$ to $_{12}C$, but as the $_{14}C$ decays with no possibility of replenishment, the proportion of carbon 14 decreases at a known constant rate. Thus, the measurement of the remaining proportion of $_{14}C$ in organic matter gives an estimate of its age (a raw radiocarbon age).

Decay of Carbon-14

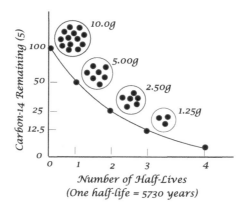

Number of Half-Lives
(One half-life = 5730 years)

The equation governing the decay of a radioactive isotope is

$$N = N_0 e^{-\lambda t},$$

where N_0 is the number of atoms of the isotope in the original sample (at time $t = 0$) and N is the number of atoms left after time t. λ is a constant that depends on the particular isotope; for a given isotope it is equal to the reciprocal of the average or expected time a given atom will survive before undergoing radioactive decay. The mean-life, denoted by τ, of $_{14}C$ is 8,267 years, so the equation above can be rewritten as

$$t = 8267 \ln \frac{N}{N_0}.$$

The ratio of $_{14}C$ atoms in the original sample, N_0, is taken to be the same as the ratio in the biosphere. So measuring N, the number of $_{14}C$ atoms currently in the sample, allows the calculation of t, the age of the sample.

Willard Libby published a paper in 1946 in which he proposed that the carbon in living matter might include carbon-14 as well as nonradioactive carbon. Libby and several collaborators proceeded to experiment with methane collected from sewage works in Baltimore, demonstrating that it contained radioactive carbon-14. By contrast, the methane created from petroleum had no radiocarbon activity. The results were summarized in a paper in the journal *Science* in 1947, in which it was pointed out that it would be possible to date materials of organic origin containing carbon. The proposed dating method was cross-checked with other methods and the results were published in *Science* in 1949. This marked the "radiocarbon revolution" in archaeology, and soon led to dramatic changes in scholarly chronologies. In 1960, Libby was awarded the Nobel Prize in Chemistry for this work.

The radiocarbon dating method involves several assumptions: for example, that the level of $_{14}C$ in the biosphere has remained constant over time. In fact the level of $_{14}C$ in the biosphere has varied significantly, and as a result the values provided by the equation above have to be corrected by using data from other sources. The calibration required is about 3%.

7.17 Molecular Design

The Landscape of Molecular Design

The cost of pharmaceutical research and development (R & D) has increased dramatically over the past two decades while the number of new drug applications approved has been relatively flat. Innovative and cost-effective approaches to drug discovery are essential for any pharmaceutical company wishing to stay in the game and be competitive. Molecular design is a new promising field that uses computer and chemical compound databases to screen, model, and design *new molecular entities* (NMEs), all in order to accelerate the discovery process and reduce its costs. Why can computers can help in this aspect? One reason is that the supply of compounds is actually somewhat limited. Many companies use blind screening for the same *compound libraries* even after many of the compounds have been proved structurally unfavorable. Another reason is that the traditional biochemical screening assays can be very expensive, especially for *cell-based systems* or when elaborate protein purification schemes are required.

Molecular Design for New Drugs

Over 10 million nonredundant chemical structures cover the actual chemical space, out of which only about a thousand are currently approved as drugs. A key to drug discovery is to find a *ligand* (chemical compound) that geometrically fits the protein target just like finding a key that fits a lock. This is because in order for a drug to work it must fit the protein closely so that it can interact with its target. Chemogenomics is the new interdisciplinary field that attempts to fully match the target (protein) and ligand (chemical) spaces, and ultimately identify all ligands of all targets.

The Traditional Approach to Drug Discovery

Pharmaceutical companies customarily purchase a large number of compounds from third party suppliers and synthesize chemically diverse combinatorial libraries of compounds. The synthesized compounds will cover some sufficient portion of the entire chemical space. For each compound a screening test is performed, usually a simple binding of assays, to determine if the compound binds to the target protein. A "hit" means that an interaction between the compound and the protein is found. There are two important stages in the process: lead generation and optimization. *Lead generation* is the process of finding if the compound binds well, whereas *lead optimization* is the process of discovering how to make the compound bind better. Chemical similarity is critical at both stages.

When a "lead" has been identified, the next step is to find compounds that are similar to it, which might bind even better. This can be accomplished through similarity searching in an existing compound library. Chemists can also make specific changes to the lead compound to improve its binding affinity and other properties.

The steps in the standard drug lead-approach include (1) identifying the target (e.g., enzyme, receptor, ion channel, transporter), (2) determining DNA structure and/or protein sequences, (3) elucidating structures and functions of proteins, (4) proving the therapeutic concept in animals ("knock-outs"), (5) developing assays for high-throughput screening, (6) mass screening, and (7) selecting lead structures.

The Innovative Approach to Drug Discovery

A goal of molecular design is to identify novel substances that exhibit desired properties such as a particular biological activity profile that includes selective binding to a single target or desired activity modulation of multiple targets simultaneously. This design certainly must include proper physicochemical and ADMET (absorption, distribution, metabolism, excretion, and toxicity) properties of the novel compounds (Schneider and Baringhaus, 2008).

A binding interaction between a small molecule ligand and an enzyme protein may result in activation or inhibition of the enzyme. If the protein is a receptor, ligand binding may result in *agonism* (activating a *receptor*) or *antagonism* (blocking a receptor from activation). Docking, a procedure to identify a ligand matching a protein target, is most commonly used in drug design. The docking can be used for (1) virtual screening or hit identification, i.e., to quickly screen large databases of potential drugs in silico to identify molecules that are likely to bind to protein target of interest, and (2) lead optimization, i.e., predicting where and in which relative orientation a ligand binds to a protein, which is called the binding mode or pose.

The design of ligands that interact with a target receptor always yields a low energy ligand–receptor complex. This ligand conformation is not necessarily the global minimum-energy conformation.

The term *drug-like* describes various empirically found structural characteristics of molecular agents, which are associated with pharmacological activities. It is not strictly defined but provides a general concept of what makes a drug a drug. Drug-likeness may be considered as an overall likelihood that a molecular agent can be turned into a drug. Therefore, it is a measure of complex physicochemical and structural features of the compound, which can be used to guide the rational design of lead structures that can be further optimized to become a drug. The components of such a *structure-activity relationship* (SAR) can include (1) substructure elements, (2) lipophilicity (hydrophobicity), (3) electronic distribution, (4) hydrogen-bonding characteristics, (5) molecule size and flexibility, or (6) pharmacophore features (Sadowski & Kubinyi, 1998; Schneider and Baringhaus, 2008).

Once an SAR model is available, it is possible to perform rational drug design (Höltje et al. 2008). SimScore (Sakamoto et al., 2007) was developed to evaluate quantitatively the structural similarity score of a target compound with the teratogenic drugs which are defined as serious human teratogens by the FDA. In SimScore, a molecular structure is divided into its skeletal and substituent parts in order to perform similarity comparison for these parts independently. The principle behind SimScore is that compounds with the same or similar skeletons will show a similar biological activity, but their activity strengths depend on the variation of substituents.

Biochemical processes within a cell or an organism are generally caused by molecular interaction and recognition events. It is critical to identify disease pathways and discover molecular structures that can be used to interfere with and slow down disease progression, or that may eventually cure the disease.

Receptor-ligand complexes between drug molecules and their macromolecular targets are mostly based on noncovalent interactions. Covalent binders exist, but typically, reversible binding causing reversible pharmacological effects is desirable, as the covalent binding of drugs is often eventually responsible for various types of drug toxicity.

A well-known example of the application of structure-based drug design leading to an approved drug is the carbonic anhydrase inhibitor dorzolamide, which was approved in 1995 (Erickson, Baldwin, and Varney, 1994; Gubernator and Böhm, 1998). Another important case is imatinib, a tyrosine kinase inhibitor. Imatinib is substantially different from previous drugs for cancer, as most agents of chemotherapy simply target rapidly dividing cells, not differentiating between cancer cells and other tissues.

7.18 Clinical Trials

As we have mentioned earlier, clinical trials are divided into chronological phases. The size of a trial is increased only gradually from Phase 1 to Phase 3 for reasons of patient safety and cost reduction. Upon the completion of Phase-3 trials, efficacy and safety results are presented in a so-called *integrated efficacy summary* (ISE) and *integrated safety summary* (ISS). These integrated results as well as other documents are organized according to ICH guidance to submit to the *regulatory agency* for approval. In the United States, after a 10-month review process, the sponsor will receive a response from the FDA regarding their new drug application (NDA). There are three types of FDA response: (1) Approval for marketing the drug, (2) application denied, and (3) incomplete response/request for more information.

The clinical trial processes typically include protocol development, investigator (or clinic site) selection, patient recruitment, administration of the drug and measurement of outcome per protocol, data entry and validation, statistical analysis, study report writing and publications, and regulatory filing as necessary.

For ethical reasons, patients are required to sign the informed consent form (ICF) before they can enroll to the study. An ICF is a document summarizing the trial's purpose, the potential benefits and risks, and the privacy policy.

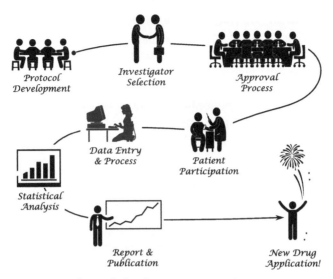

Protocol Development — *Investigator Selection* — *Approval Process* — *Data Entry & Process* — *Patient Participation* — *Statistical Analysis* — *Report & Publication* — *New Drug Application!*

Clinical Trial Process Overview

Protocol development is a very important first step in any clinical trial. A clinical trial protocol, developed by the sponsor and approved by an investigational review board (an external experts panel) and the FDA, is a document about the trial's design and conduct. To protect trial validity and integrity, Good Clinical Practice (GCP) requires that all study investigators should adhere to the protocol in conducting the clinical trial.

The *protocol* describes the scientific rationale, objective(s), endpoints for efficacy and safety evaluations, test drug (and its competitor if any), dose regimen, randomization, assessment schedule, data collection, size of the trial, and statistical considerations. It contains a precise study plan for executing the clinical trial, not only to assure the safety and health of the trial subjects, but also to provide an exact template for investigators so that the study will be performed in a consistent way. This harmonization allows data to be combined from all investigators. The protocol also gives the study monitors and the site team of physicians, nurses, and clinic administrators a common reference document for site responsibilities during the trial.

The protocol typically includes elements that we will illustrate with a hypothetical drug for treating patients with iron deficiency anemia.

(1) *Study Objective*

The purpose of this study is to evaluate the efficacy and safety of an injection drug MyIron, compared with placebo (normal saline) for the treatment of iron deficiency anemia (IDA).

(2) *Target Population*

The target population is the large population for which the test drug is intended to treat if approved for market. It is fundamentally important for patients and for the company. On one hand, the test drug may not work for a very broad population. On the other hand, narrowing the population can reduce the number of patients who potentially get the benefits, thus reducing the profit of the drug maker. Therefore, an appropriate choice of target population is critical. The target population can be precisely defined by inclusion and exclusion criteria. Patients who meet all inclusion criteria and none of the exclusion criteria can enroll into the trial.

For this IDA study the key inclusion criteria are

- IDA with a screening hemoglobin (Hg) value <10.0 g/dL
- Patients who couldn't take oral iron
- Male and female subjects $\geq$ 18 years of age

The key exclusion criterion is erythropoiesis-stimulating agent (ESA) therapy used within 4 weeks prior to screening. This is because ESA will dramatically affect the efficacy outcome and contaminate the efficacy evaluation. The 4-week washout period is to ensure the previous ESA effect is eliminated and the hemoglobin value comes back to the "baseline".

(3) *Study Design*

This is a Phase-3, randomized, double-blind, placebo-controlled, multicenter clinical trial to evaluate the safety and efficacy of the study drug compared with placebo for the treatment of IDA.

The randomization is to minimize the imbalance between the treatments with respect to the confounding factors. The double-blindness means that the patients and clinicians do not know which treatment (MyIron or placebo) is given to which patient to avoid an affirmative bias. The placebo is to avoid the bias caused by the placebo effect. In other words, this placebo effect will be subtracted from the total observed effect in the experimental group because the total effect includes both the true effect of the drug and the placebo effect.

Total subject participation is approximately 7 weeks, which includes a 2-week Screening Period and a 5-week Treatment Period. Up to 3 screenings can be performed on each patient to meet the inclusion criterion, Hg<10.0g/gL.

(4) *Endpoints*

The primary endpoint is the key measure of efficacy outcome in the trial, indicating whether the test drug works. The primary endpoint must be clinically meaningful and sensitive enough so that its changes can be easily detected. Iron deficiency can be reflected in a low Hg value. MyIron is expected to increase the Hg value. Thus, for this study, the primary endpoint is the proportion of subjects achieving a $\geq$2.0 g/dL increase in hemoglobin at any time from Baseline to Week 5 of the treatment period. The secondary endpoint is used to support the efficacy claim of the test drug. People with IDA often appear fatigued. Thus, the secondary endpoint is defined as the mean change in Functional Assessment of Chronic Illness Therapy (FACIT)—the Fatigue score from Baseline to Week 5

(5) *Randomization*

Randomization into treatment groups will be stratified by baseline hemoglobin level (>7.0 to ≤8.5 g/dL; >8.5 to <10.0 g/dL). Subjects will be assigned to either MyIron or placebo in a 1:1 ratio within each of the strata. Such a stratified randomization can provide a high probability of balance between the treatment groups with regarding to the important (potential) confounding factors.

(6) *Sample Size*

Based on the primary efficacy endpoint of the proportion of subjects who achieve a ≥2.0 g/dL increase in hemoglobin at any time from Baseline to Week 5, a sample size of 600 subjects (300 exposed to MyIron and 300 exposed to placebo) provides a more than 95% power for the assessment of superiority of MyIron to placebo.

7.19 Incredible Power of Little Statistics

In many practical problems, you don't need a complex theory of statistics, but some simple knowledge of statistics will do the job for you as illustrated in the following four examples.

(1) *Accounting Fraud Detected by Benford's Law*

Benford's law, though derived from empirical observations, is counterintuitive. It says that in lists of numbers from many real-life sources of data (e.g., electricity bills, stock prices, population numbers, death rates, lengths of rivers, physical and mathematical constants), the leading digit is, surprisingly, not distributed uniformly. In fact, the distribution, like other physical laws, is independent of the units chosen (e.g., meter or inch). *Benford's law* states precisely that the leading digit d ($d = 1, 2, \ldots, b - 1$) in base b ($b > 2$) occurs with probability (Benford, 1938)

$$\Pr(d) = \log_b\left(1 + \frac{1}{d}\right)$$

Benford's law has been used in fraud detection. As early as 1972, Hal Varian suggested that the law could be used to detect possible fraud in lists of socio-economic data submitted in support of public planning decisions. Based on the plausible assumption that people who make up figures tend to distribute their digits fairly uniformly, a simple comparison of first-digit frequency distribution from the data with the expected distribution from Benford's law ought to highlight any anomalous results. Following this idea, Mark Nigrini (1999) showed that Benford's law could be used as an indicator of

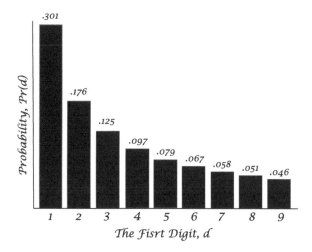

Benford's Law with Base $b = 10$

accounting and expense fraud. In the United States, evidence based on Benford's law is legally admissible in criminal cases at the federal, state, and local levels. Benford's law was invoked as evidence of fraud in the 2009 Iranian elections (Battersby, 2009). Benford's law itself is usually not sufficient to determine an account fraud, but it can be used as an indicator for the necessity of a further investigation.

(2) *Discovering the Total*

In Section 7.10, we use the Mark–Recapture Methods to calculate the size of a population. Here is another type of total-finding problem. Suppose school buses are numbered sequentially from 1 to N. One day you see a bus numbered K (e.g., 8) passing by. Curiously, you asked yourself: how many school buses are there in the town?

At first glance, you may have thought that there was no way to know it from this single number. But actually, you can estimate the total number of buses by $N = 2K - 1$. This is because you can see a big or small bus number, but the mean of the bus numbers you see should be $K = (1+N)/2$. (a question to the reader: how do you determine whether the estimated number is for the total number in a town, a state, or a country?). Of course, this estimate is very inaccurate because you have seen only one bus number. If you see more bus numbers, the mean can be more accurately estimated as well as the total, N. We illustrate this with the following example.

Suppose a business investor wants to know the total sales of a competitive product. He knows that each shipped product has a unique serial number. The serial numbers consist of continuous 6-digit numbers, starting with the number 000001. How can he obtain the information about the total sales of the competitive product? It could be a couple of hundreds or it could be close to a million. Here is a simple way to estimate the total sales: randomly get some samples of the product that have been sold. Assume their serial numbers are 003002, 001023, and 007234. The mean is $(3002 + 1023 + 7234)/3 = 3753$. The total sales is estimated to be $2(3753) - 1 = 7505$. But what if we don't know the starting number? How do we solve the problem in such a case? Here is a simple way.

The *mean deviation* (MD) is the mean of the absolute deviations of data about the mean. For a large continuous sequence of N numbers (regardless of the starting number), the expected MD is about $N/4$. Therefore, the total is $N = 4MD$. MD can be easily calculated from the data. For the above example, the MD is $(|3002 - 3753| + |1023 - 3753| + |7234 - 3753|)/3 = 2320.7$. Thus, the total sales is about $4(2320.7) = 9283$. The result is somewhat different from 7505 obtained from the previous method, but it is not surprising that different methods give different answers. Fortunately, when the sample size gets larger, the two solutions will become closer and more accurate.

(3) *Drug Effect Revealed from Blinded Data*

We know that *variance* is the second central moment of a distribution, which measures the variability of the data. *Skewness* is the third central moment, measuring the asymmetry of the distribution. For normal distribution the skewness is zero because of the symmetrical distribution. *Kurtosis* (standardized) is the fourth central moment, measuring the flatness of the distribution. For normal distribution, the standardized kurtosis is 3. In general the nth central moment of a random variable X is defined as the expectation of $(X - \mu)^n$, where μ is the population mean. These moments, despite their simplicity, can be very useful. For example, it is common to believe that we cannot estimate the treatment difference between the test drug and placebo based on the pooled (treatment blinded) data. But that is not exactly true. The treatment difference can be calculated without separating the two treatment groups. This is because data from the mixed distribution of two different normal distributions (not the distribution of sum of two normal variables) will have two modes if the difference between the two means is sufficiently large and a single mode if the two means are the same or close to each other. In general the mean difference is reflected in the variance and kurtosis. Let me elaborate on how we can calculate the mean difference from the pooled data.

Suppose we have two equally sized groups of patients, treated with either placebo or the test drug. Assume the treatment effects are normally distributed with means μ_1 and μ_2, respectively, and a common variance $\tilde{\sigma}^2$. The variance of the mixed distribution is

$$\sigma^2 = \tilde{\sigma}^2 + \delta^2,$$

where δ is a half of the treatment difference; that is, $\delta = |\mu_2 - \mu_1|/2$. This formulation implies that the variance calculated from the pooled data is δ^2

Mean Difference Increases

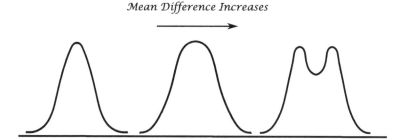

Mixed Distribution of Two Normal Variables

larger than the variance calculated for each group separately. The kurtosis κ is

$$\kappa = \frac{1}{\sigma^4}\left(3\tilde{\sigma}^4 + 6\tilde{\sigma}^2\delta^2 + \delta^4\right)$$

From these two equations we can solve for δ,

$$\delta = \sigma\left(\frac{3-\kappa}{2}\right)^{1/4}.$$

The σ and κ can be estimated or calculated from the pooled data, and thus, the δ is obtained. The accuracy of the calculated δ will depend on the accuracy of the calculated σ^2 and κ, which in turn depend on the actual treatment difference and the sample size. For a larger treatment difference ($2\delta > 1$), a very small sample size is sufficient; for a smaller difference ($2\delta < 0.2$), a large sample size (1000 or more) is required.

(4) *Strategy of Choice*

In our lifetime, we have to make many critical decisions and choices: choosing an outfit for the day, deciding on the best school to attend, buying the right car, going to the best place for vacation, finding the best job, marrying the person we love the most, etc. We may face two distinct situations: (1) All available options are presented to the decision maker at the same time. This is a relatively easy problem but it may be time-consuming to pick the best one when the number of options is larger. (2) The options are presented sequentially over time. At a given time, only one or a few options are available to the decision maker. In this case, how should we make a choice?

The most popular Chinese TV show, "FeiCheng WuRao," is a dating show where there are 24 unmarried women and one male in each "game show." Female dating participants can see only one male candidate at a time. Thus, the females don't know what type of male to expect for the next round: he could be a better man than the present one or could be a worse one. If a female candidate decides not to date the current male candidate, she will turn off her light and lose the opportunity permanently in the current game. Each male gets to pick one female from all female candidates who decide to keep their lights on at the end of the game. Suppose this is the only place for a female candidate to meet someone she will love, she has only one chance to select, and she decides to see at most $n = 100$ male candidates. What strategy should she use in order to have a good chance to date the best one? If she randomly picks one, the chance of picking the best one is $1/n$, which diminishes when n gets larger and larger. Fortunately, there is a simple strategy that will give her nearly a 40% chance of picking the best one (Chang 2012, Székely 1986).

Here is what she needs to follow: Let the first 37% (more precisely, $100/e\%$) of the candidates go and then select the first one better than any previous candidate (if none is better, select the last). In this case, the chance of selecting the best is $1/e \approx 37\%$, regardless of the value of n. In reality, the n is often not exactly known; therefore, it is difficult to know when the 37% candidates have gone by. But fortunately, even if it is not 37% of candidates, the probability of choosing the best is still reasonably good as you can see from the computer simulation results in the following table. If you want the expected rank of the selected candidate to be the best, you would let the first 10% (instead of 37%) of candidates go. The best candidate is ranked 1 and the worst candidate is ranked 100 in the table.

Probability of Selecting Best Candidate

Let-go Candidates (%)	5	10	20	30	37	50	60	70	80	90
Probability (%)	15	24	33	36	37	35	31	25	18	10
Average Rank (%)	11	10	12	17	20	26	31	36	41	46

7.20 Publication Bias

Publication bias is the phenomenon that more findings in scientific publications contain positive results than are actually warranted. As mentioned in Section 3.8, some publication bias can be avoided, while some cannot. Let's look into the statistical nature of publication bias. There are three different sources of publication bias that are not caused by researchers' bad practices: positive reporting bias, regression to the mean, and multiplicity.

Given that there is no intentional manipulation of data or results or misuse of any analytical methods, if we report all experimental outcomes (negative or positive), there will be no bias, meaning there are equal numbers (and magnitude) of false positive and false negative findings. Now, if we select only the positive outcomes (e.g., drugs that show abilities to cure disease), such a subset of all outcomes will naturally have more (and larger) false positive findings than false negative findings. I call such bias positive reporting bias.

The second statistical nature of publication bias is "regression to the mean" (Section 6.7). Regression to the mean can be simply stated as: the top performers the first time around will likely perform worse the second time. If we perform the second studies as validation for the positive findings in the first studies, we will likely find that the second studies have a smaller magnitude or a smaller number of positive findings. When these "verified positive findings" are further validated with third studies, the positive findings will likely further reduce due to, again, regression to the mean. As more verification processes are undergone, false positive findings will be gradually eliminated, but at the same time we lose some true positive findings (type-II error) each time. Eventually, both the false and true positive findings will vanish.

The third nature of publication bias is due to multiplicity (Section 6.9). The current drug approval rate in U.S. is about 40%, i.e., 60% of Phase-3 trials fail, which adds a huge cost to drug development. Suppose 50% of drugs in Phase-3 are not efficacious (no better performance than the drugs on the market), and say 1000 ineffective compounds are in Phase-3 trials each year in the pharmaceutical world, then 25 of them will have false positive findings, i.e., be falsely proved to be efficacious (one-sided test at 2.5% level of significance), and 20 will have false negative findings (if the actual power is 80%) among the 1000 true positive drug candidates. Similarly, if there are 100 researchers studying the effect of the same genetic marker on a disease, one of them will likely discover the relationship between the marker and the disease even if there is actually no such relationship, and four of the researchers on average will confirm the false finding (based on a two-sided test at 5% level). Such multiplicity conducted by different companies or researchers using different experimental data has not been controlled at all. The exploratory nature of science puts the validity of research findings in questioning, even if the p-value shows less than α, the level of significance. The reason is that many researchers have tried many different statistical models, hypotheses, and data cuts to innocently "fish statistical significance" without fully realizing the multiplicity issue. Multiplicity can become worse before getting better if the data become more accessible to the public—since everyone can use the same data or portion of the data to "discover" patterns or test new hypotheses.

Multiplicity in meta-analysis is particularly problematic, but understandable. Data are usually not uniform in experiment design, combinations of data are somewhat arbitrary, many people are conducting different analyses with different hypotheses with slightly different or the same combinations of data, and the analyses are post hoc since hypotheses are proposed after each study's dataset has been analyzed. Having said that, meta-analyses do have the advantage of having a larger sample size.

Now the question is: How severe is publication bias? John Ioannidis, Professor of Medicine and Director of the Stanford Prevention Research Center, studied publication bias and published an article on the topic in JAMA

(2005a). Among 49 highly cited (>1000 times) clinical research papers, he compared the 45 studies that claimed to have uncovered effective interventions with data from subsequent studies with larger sample sizes. Seven (16%) of the studies were contradicted, and for 7 (16%) the effects were smaller than in the initial study. Thirty one (68%) studies remained either unchallenged or the findings could be replicated. From these data alone (not combining with a low prior), we cannot see any evidence to support a claim he made in a later article that most published research findings are false; on first observation one notices that the sample size 45 is pretty small compared with the huge number of publications in the medical fields, and false findings in general scientific journals may be higher.

Experimental designs in the medical field are often powered at 80% to 90% based on estimated effect sizes. However, due to positive bias in estimating the effect size, the true effect size is usually smaller. Therefore, the actual power may be around 70% (or lower) with a range from 5% to 95%. If a trial is actually powered at 70%, then among 45 true positive findings, 0.7(45) = 31.5 findings are expected to be validated (reproducibility), which matches Ioannidis's findings of 31 studies with unchanged findings. In other words, among those highly cited papers, the positive findings are virtually all truly positive. If there are only 23 (still more than 50%) true positive findings that are false positive, then only 20 studies can show positive, on average, in the 45 follow-up validation studies. If we believe the 45 follow-up studies (realizing the small sample size) are good studies and Ioannidis did a good job in his paper, then we are confident that a majority of the positive findings in highly cited papers are true positive, with virtually no false positive findings.

It is interesting that a month after his JAMA paper, Ioannidis (2005a), published another groundbreaking article entitled "Why Most Published Research Findings Are False" (2005b). It seems the two papers from the same author conflict each other. In this paper, he starts with a sentence "...the probability that a research finding is indeed true depends on the prior probability of it being true (before doing the study). ..." Goodman and Greenland (2007a, 2007b) rejected the claim of Ioannidis that most published research findings are false by pointing out: "Unfortunately, while we agree that there are more false claims than many would suspect—based both on poor study design, misinterpretation of p-values, and perhaps analytic manipulation—the mathematical argument in the *PLoS Medicine* paper underlying the 'proof' of the title's claim has a degree of circularity." In other words, by Bayesian approach, it is concluded that

$$\text{strong negative prior} + \text{positive data} = \text{mild negative result}$$

Wren (2005) puts it this way: "... his claim that most published research findings are false is somewhat paradoxical. Ironically, the truer his premise is, the less likely his conclusions are. He, after all, relies heavily on other studies to support his premise, so if most (i.e., greater than 50%) of his cited studies are themselves false ..., then his argument is automatically on shaky ground (e361)."

Most people would probably be surprised when first reading that "most published research findings are false," which means that they probably have a positive prior (less than 50% false findings in publication) before seeing Ioannidis's paper. Therefore, for them the conclusion would be

positive prior + positive data = positive result.

Thus, I accept as credible Ioannidis's claim no more than the findings by his peers. From that we have the Liars' Paradox in a probabilistic form: If Ioannidis is correct and, say, 70% publications are false findings, by applying this to the statement itself (Ioannidis's claim), we conclude there is a 70% chance that Ioannidis's claim is wrong. This, in turn, means the majority of the publication's findings are true. That, in turn, implies Ioannidis's finding is probably true. The insidious process goes on.

Despite Ioannidis's good intentions, his claim "Most published findings are false" is not supported by the evidence so far. His statement might be true and might be just a false finding fighting against false findings. I am concerned about false findings in populations due to innocent misuses of statistical methods since prominent scientists such as Ioannidis did not use statistical methods rigorously or without a statistician's assistance (Goodman and Greenland, 2007a, 2007b) in his paper "Why Most Published Research Findings Are False" (2005b). Ironically, the paper has been the most accessed article in the history of the Public Library of Science (approximately 1 million hits, see Ioannidis's CV); if it is not for the title, there must be something else striking the public's curiosity.

As researchers, what we can do to reduce publication bias? Increasing the power of experiments will reduce the bias but it will be costly and may prevent us from spending resources wisely. Increasing confirmative experiments can remove most of the false positive findings, but can be costly and will remove some true positive findings (i.e., increase false negatives) at the same time. Finding a balance between exploratory and confirmatory studies is an efficient way to minimize false positive findings (Section 3.11).

Raising the bar for positive claims (e.g., setting the significance level α at 1% instead of 5%) can usually reduce the proportion of false positive findings among all positive findings (i.e., the false discovery rate), but at the same time it is often costly because a larger sample size may be needed to retain the same power. Besides, raising the bar for positive claims alone without increasing sample size does not necessarily reduce the magnitude of bias. This is because, given the true effectiveness of a drug a higher bar will lead to selecting more positive observed results to publish, thus more bias from the true effectiveness. For example, if 100 observations X are drawn randomly from N(0.5, 1)... will disregard a larger portion of "negative value" than using a lower threshold, the mean of $x > 0.196$.

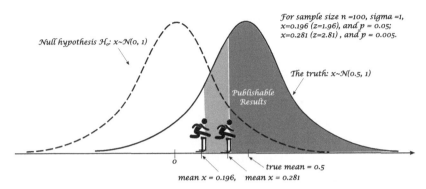

We usually publish the result (finding) when it is statistically significant, i.e., $p<0.05$, or equivalently mean $x > 0.196$ in this case. However, to tighten false positive findings, one may choose to publish the results only when $p<0.005$, or mean $x > 0.281$, knowing that it will have a larger publication bias.

Reducing the Error Rate (Significance level) Can Increase the Bias of the Published Mean

Increasing the hurdles for positive claims does not necessarily improve the false positive rate if, for example, it results in people studying more compounds with smaller sample sizes instead of fewer larger studies. Thus, instead of running a 400-patient trial to test a single compound, one can use the same sample size to run 200 trials with 2 patients per trial to test 200 different compounds. Even when none of the 200 compounds is effective, one will be able (given a corresponding one-sided alpha of 0.025) to claim 5 compounds effective on average, as discussed in the Spring Water Paradox in my previous book (Chang, 2012). Luckily, p-value is just one basic requirement for approval of a drug. There are many other requirements and a very small sample will not be allowed in Phase-3 clinical trials.

Mistakes such as publication bias can also occur in other situations. For instance, I saw a publication a few years ago accusing pharmaceutical companies of making too much profit. The authors' conclusion is based on their investigation of pharmaceutical companies which have successfully made drugs or on the entire cost of the companies' successful drugs. But their result was guilty of "positive reporting bias" because many failed pharmaceutical/biotech companies that have filed bankruptcies were not included in their cost base. In an analogy, one cannot complain about a lottery winner taking too much profit without considering the larger number of losers. The pharmaceutical industry is a high risk business entity: for their successes, 50% depends on good research and the other half depends on good luck.

Ioannidis published a third article in Scientific American (May 31, 2011) entitled "An Epidemic of False Claims—Competition and Conflicts of Interest Distort Too Many Medical Findings." I just want to remind readers that most publication bias is due to the nature of scientific research (positive reporting bias, regression to the mean, and multiplicity). Most scientists perform their diligent work with the high ethical standard (Section 3.11) and make great contributions to the advancement of science. Ioannidis makes several good suggestions to prevent bias caused by malpractice. One interesting suggestion he has made is to report the number of analyses that have been done in order to assess type-I error. But that is not always possible since, for example, for meta-analyses (one of Ioannidis's two greatest research interests), how does a researcher know how many analyses have been done by other researchers on the same data, and thus how can he avoid a large number of false findings? I hope that Ioannidis's paper raises the level of the already existing public attention to the issue of publication bias and promotes an honest scientific attitude, but does not mislead the public that a majority of the false findings in the publications are caused by unfaithful practice. I work in a highly regulated environment and I know the work performed by my colleagues is held to extremely high scientific standards. What I don't want to see is that scientists are wrongfully accused because of a misuse of statistics by the accuser. For that, I strongly recommend statisticians should be involved with this aspect of their research. They should take responsibility, as a safeguard to minimize false positive findings in publications, by removing innocent misuses of statistical methods, especially those involving multiplicity issues.

We should look at both sides of research dynamics: (1) competition, on one hand, can distort medical findings; on the other hand, it can raise the quality of research; (2) an author's research degrees of freedom, especially in meta-analysis, on one hand increases the type-I error; on the other hand more exploratory analyses will bring the true best findings to the confirmatory research; (3) a reduction in false positive findings often implies an increase in false negative findings, which is also a waste of resources and time; (4) research technology and methodology development can improve the quality of papers.

If there is no false discovery, there would be no true positive discovery, no science. Given the large number of exploratory analyses in the publications, the possibility that most findings are false might be an inevitable part of the research endeavor (*PLoS* Editors, 2005). We want to reduce the proportion of errors, but without slowing down the discovery of the true interconnections present in nature. A central issue to this challenge is the multiplicity issue: the authors' degree of freedom in research or analysis, the freedom to try different modes for fitting the same data, freedom to do different meta-analyses on the same data again and again, freedom to test a hypothesis repeatedly with different experimental data. We don't have a perfect solution yet, but raising the awareness of multiplicity issues and finding the right balance between exploratory and confirmatory studies seem the direction to follow in solving the problems.

7.21 Information and Entropy

In the 1940s many scientists, mathematicians, and engineers toiled in war-related fields. Working both in groups and individually, goals were set and progress was made on the serious and practical problems of coding, code-breaking, communication, and automatic control. Among the scientists at the forefront of these endeavors were two Americans. Claude Shannon worked out the discrete case of standard entropy/coding/communication channels as part of information theory at the same time as Norbert Wiener was doing the continuous version.

In his 1948 paper "A Mathematical Theory of Communication," Shannon, using a narrow definition of information, proved a very important theorem, which gave the maximum possible transmission rate of information over a given channel (a wire or other medium). This maximum transmission rate is called the channel capacity. Since its inception, information theory has broadened to find applications in many other areas, including natural language processing, neurobiology, molecular codes, thermal physics, quantum computing, and many others. Applications of the fundamental topics of information theory include lossless data compression (e.g., ZIP files), lossy data compression (e.g., MP3s and JPGs), and channel coding (e.g., for Digital Subscriber Lines (DSLs)). The field lies at the busy intersection of mathematics, statistics, computer science, physics, neurobiology, and electrical engineering. Its impact has been crucial to the invention and success of the compact disc, the feasibility of mobile phones, the development of the Internet, the study of linguistics and human perception, as well as the understanding of black holes, and numerous other physical processes.

The maximum possible transmission rate of information relates to the concept of *entropy*, which is a measure of disorder. To understand this, let's start with a number guessing game. Imagine that your friend hands you a sealed envelope. He informs you that it contains an integer from 1 to 10, and asks you to guess what it is. However, first you get to ask yes-or-no questions about the unknown number, to which he'll respond truthfully. Assuming this game is repeated many times over, and you become as clever at choosing your questions as possible, what is the smallest number of questions needed, on average, to pin down the number?

Information and a Guessing Game

You use the so-called binary search, first splitting the ten numbers into two equal sets, those less than 6 and the rest. So you ask: "Is it less than 6?" He answers either "Yes" or "No." Assume he answers "No." Knowing now that the integer is between 6 and 10 you divide the smaller set of numbers into two equal parts and ask "Is it less than 8?" Suppose his answer is "No," so that the number is either 8, 9, or 10. You continue, "Less than or equal to 9?" He says: "Yes." This is your fourth question: "Is it 9" He says: "No." Finally you have the number, 8. But this is the worst case scenario. You have been unlucky: if the answer to the third question had been "No," you would have known the number at that moment. Instead it has taken you four yes-no questions before determining the integer's value. Clearly, four is also the minimum number of yes-no questions (wisely chosen) that you need to ask your truthful friend to be sure of nailing down the answer. This minimum can actually be calculated from the equation $2^n = 10$ or $n = \log_2 10 = 3.3 \approx 4$. If the unknown number is between 1 and 100, the number of questions needed is $n = \log_2 100 = 6.6 \approx 7$. The value of n increases slowly, $n = \log_2 1000 \approx 10$. In general, the minimum number of yes-no questions required to determine with certainty a number between 1 to N is $n = \log_2 N$.

Guessing a number is a pretty boring thing to repeat. Let's guess a message with K English characters. First, we know there are $N = 27^K$ possible K-character strings since English has 26 characters and the white space. Let's code these strings, i.e., map each string to an integer from 1 to N. Using the coding, guessing the message is no different from guessing a number because we can always ask, for example, "Is the coding number of the message less than 8?" For 5-character messages, K is equal to 5 and the number of guesses is $\log_2 27^5 \approx 24$. This is the worst case scenario; you may be lucky in getting a message requiring fewer questions (guesses). On average the number of questions needed to get the correct number is proportional to $\log_2 N$ or equal to $K \log_2 N$, where K is a constant. Because information is measured in a relative sense, the constant K can be ignored ($K = 1$).

So far we have assumed the characters in the message are in perfect disorder and meaningless. In English (and other languages), not all characters or sequence characters are equally possible. For example, there are no such words: aaaaa or zzzbb and there are grammars to follow. We denote the probability of the message code $= i$ by $\Pr(S = i) = p_i$. Using this probability, the average number of yes-no questions needed is proportional to

$$- \sum_i p_i \log_2 p_i.$$

This gives us $\log_2 N$ when all the p_i are equal. The sum is called the Shannon entropy, conventionally written $H[S]$.

We should make it clear that entropy of a source S is a measure of disorder. It is also the maximum possible amount of information that can be stored in S. By following certain rules, e.g., English grammar, the information possibly contained in a source is reduced. Entropy is the amount of information that is unpredictable, which excludes the average amount of information that can be predicted by English grammar.

Now we may wonder what additional information we gain by knowing sources T and S, given that we already know T. Equivalently we can ask: What is the difference in entropy between T alone and the combination of S and T? That is, we calculate

$$H[S|T] = H[S, T] - H[T].$$

Here the entropy of the combination of S and T, denoted by $H[S, T]$, is at most $H[S] + H[T]$ when perfect disorder (any message T can follow any message S) is present . But some combinations of messages may be more likely than others. To illustrate this, let's investigate the entropy for English as a language (Marques de Sá.,2008).

(1) The zero-order model: Randomly selecting a character based on the uniform distribution $P(S_k) \equiv 1/27$, we obtain the maximum possible entropy for our random-text system, i.e., $\log_2 27 = 4.75$ bits. (2) The first-order model: Use the probability distribution of the characters in true English texts. For instance, the rate of occurrence for the letter e is about 10.2% of the time or $P(e) = 0.102$, while the letter z has 0.04% of occurrences, so $P(z) = 0.0004$, etc. Using this first-order model, the entropy decreases to 4.1 bits, reflecting a lesser degree of disorder. (3) The second order model: If we impose true probabilistic rules between two conjunctive letters in English, the entropy for this model of conditional probability, known as conditional entropy, is computed in the following way:

H[letter 1 if letter 2] = H[letter 1 and letter 2] - H[letter 2].

Using the formulation for the calculation, the entropy of this second-order model is 3.38 bits. Moving to certain third- and fourth-order models, which introduce further restrictions on the distribution of characters, we note that the entropy for the fourth-order model is 2.06 bits.

Let n denote the rank value of the word and $\Pr(n)$ the corresponding occurrence probability. Then *Zipf's law* prescribes the following dependency of $\Pr(n)$ on n:

$$\Pr(n) = \frac{0.08}{n^{0.96}}.$$

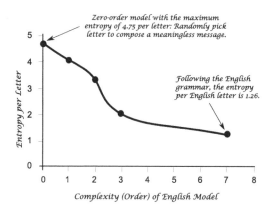

Entropy per English Letter

One may then apply the entropy formula to the value $\Pr(n)$ for sufficiently large n. An entropy of $H \approx 9.2$ bits per word is obtained. Given the average size of 7.3 letters per word in English, one finally determines a letter entropy estimate for English texts of 1.26 bits per letter. The value of entropy does vary among languages, but using this value and the download speed (in megabytes) of our internet service, we can easily calculate the maximum transmission rate of information through the cable. Practically speaking, there is noise involved in the signal transmission, so the channel capacity is also dependent on bandwidth. See Shannon–Hartley Theorem for theoretical details.

Zipf's law is empirical. Although its applicability to other types of data has been demonstrated (for instance, city populations, business volumes, and genetic descriptions), a theoretical explanation of the law remains to be found.

This is a mathematical aspect of information. But what is really going on with information and our conception of information in general? It may be something that everyone knows but, at the same time, doesn't know exactly.

It may be a kind of feeling, as Professor Alfréd Rényi (Székely, 1986, p. 223) describes:

> Since I started to deal with information theory I have often med-
> itated upon the conciseness of poems; how can a single line of verse
> contain far more 'information' than a highly concise telegram of the
> same length? The surprising richness of meaning of literary works
> seems to be in contradiction with the laws of information theory. The
> key to this paradox is, I think, the notion of "resonance." The writer
> does not merely give us information, but also plays on the strings of
> the language with such virtuosity that our mind, and even the sub-
> conscious, self-resonate. A poet can recall chains of ideas, emotions
> and memories with a well-turned word. In this sense, writing is magic.

We as information seekers are bloated by what our televisions and our mainstream media outlets give us as "news" and need to redefine our information consumption. We do not consume information deliberately and the information that we do consume is usually biased towards what we already believe. This can not only misinform us but can also waste our time and help us engrain biases that we have built up in the course of our lives. Our information diets are made up of too much entertainment and information that affirms what we already believe (mass affirmation). We consume whatever "tastes the best" and almost ignore everything else (Johnson, 2012).

Our brains are constantly overloaded with information. We often feel that less, not more information and fewer choices, is actually better. To thrive in the Information Age, we need to learn to manage our limited brain memory more effectively. In the Information Age, information, knowledge, and wisdom are finally united. However, we should fully realize that the composition of our knowledge has evolved. Functionally, An information-Age brain has two parts that carry out two distinct tasks: (1) initiating and carrying out creative undertakings, such as the classification of data and the development of new inventions, and giving voice (speech) to our ideas, and (2) the storing of metatags for classified information, while leaving intact the massive information stored in our common superbrain-Internet. Today, scientific principles across multiple disciplines serve as metatags for retrieving information, while the innate grasp of analogy remains the critical tool for human creativity and our wellspring of understanding and knowledge.

Postscript

Scientific principles are the foundation of scientific methods, holding the methods intact. Without the principles the methods would lack coherence. Science is built on the repetition of events. Since everything is unique in its own way, by grouping similar things we identify the repeating patterns of events that matter to us. On the basis of such grouping, the similarity principle is applied: similar things will behave similarly. This is the most fundamental principle of science. Only with the similarity principle can causality scientifically make sense, only on the basis of this principle can scientific laws be derived, only with this principle can we discuss inferences and predictions, only with this principle does the connotation of "scientific knowledge" become lucid, and only with this principle do we have the concept of probability and the concept of random errors. It is only because we implicitly group things differently in applying the similarity principle and form different causal spaces from which we make inferences that the theory of statistics becomes diversified into different paradigms from which controversies may arise.

Bibliography

Aggarwal, C.C. (Ed.). (2011). *Social Network Data Analytics.* Springer. New York.

Andreoni, J. and Bergstrom, T. 1993. Do government subsidies increase the private supply of public goods? The Warwick Economics Research Paper Series (TWERPS) 406, University of Warwick, Department of Economics, Coventry, UK.

Andreoni, J.A. and Miller, J.H. (1993). Rational cooperation in the finitely repeated prisoner's dilemma: Experimental evidence, *The Economic Journal.* 103(418): 570-585.

Axelrod, R. (1984). *The Evolution of Cooperation.* New York: Basic Books.

Axelrod, R. (1986). An evolutionary approach to norms. *American Political Science Review* 80(4), p. 1095–1111.

Beall, A. and Tracy, J. (2013). Women are more likely to wear red or pink at peak fertility. *Psychological Science* 24:1837–1841.

Beasley, P.R., Lin, C.C., Hwang, LY, and Chien, C.S. (1981). Hepatocellular carcinoma and hepatitis B virus. Aprospective study of 22,707 men in Taiwan. *Lancet,* 318(8256): 1129–1133.

Bellman, R. (1957a). *Dynamic Programming.* Princeton University Press, Princeton, NJ.

Bellman, R. (1957b). A Markovian decision process. *Journal of Mathematics and Mechanics* 6.

Bertrand, J. (1878). Sur l'homogénété dans les formules de physique. *Comptes rendus* 86(15): 916–920.

Beveridge, W.I.B. (1957). *The Art of Scientific Investigation.* The Blackburn Press, Caldwell, NJ.

Bickel, P.J., Hammel, E.A., and O'Connell, J.W. (1975). Sex bias in graduate admissions: Data from Berkeley. *Science* 187(4175): 398–404.

Blachman, N.M., Christensen, R., Utts, J.M., and Jinney, D.J. Letter to the editor. (1996). *The American Statistician* 50(1): 98–99.

Blyth, C.R. (1972). On Simpson's paradox and the sure-thing principle. *Journal of the American Statistical Association* 67(338): 364–366.

Bonabeau, E., Dorigo, M., and Theraulaz, G. (1999). *Swarm Intelligence—From Natural to Artificial Systems.* Oxford University Press, New York.

Boole, G. (1951). *An Investigation of the Laws of Thought.* Dover Publications, New York.

Bouton, M.E. (2007). *Learning and Behavior: A Contemporary Synthesis.* Sinauer Sunderland, MA.

Boutron, I., Estellat, C., Guittet, L., et al. (2006). Methods of blinding in reports of randomized controlled trials assessing pharmacologic treatments: A systematic review. *PLoS Med* 3(10): e425.

Bradley, F.H. (1999). On truth and copying, In Blackburn, et al. (eds.), *Truth*, pp. 31–45.

Braess, D. and Über ein (1969). Paradoxon aus der Verkehrsplanung. *Unternehmensforschung* 12: 258–268.

Braess, D., Nagurney, A., and Wakolbinger, T. (2005). On a paradox of traffic planning (translated into English). *Transportation Science* 39:446–450.

Brin, S. and Page, L. (1998). The anatomy of a large-scale hypertextual Web search engine. *Computer Networks and ISDN Systems* 30: 107–117.

Brink (2008) Psychology: A Student Friendly Approach. Unit 6: Learning. pp. 97–98

Brown, J.R. (1991). *The Laboratory of the Mind: Thought Experiment in the Natural Science.* Routledge, London.

Buckingham, E. (1914). On physically similar systems; Illustrations of the use of dimensional equations. *Physical Review* 4(4): 345–376.

Byers, W. (2007). *How Mathematicians Think.* Princetom University Press, Princeton, NJ.

Cajori, F.(1929). *A History of Mathematical Notations*, Volume II. Open Court Publishing Co., La Salle, IL.

Carey, S.S. (2012). *A Beginner's Guide to Scientific Method* (4th Ed.). Wadsworth Cenagge Learning, Boston, MA.

Cerrito, P.B. (2003). Data mining and biopharmaceutical research. In *Encyclopedia of Biopharmaceutical Statistics*, pp. 392–396. Chow, S.C (ed.). Marcel Dekker: Boca Raton, FL.

Chakravarty, A. (2005). Regulatory aspects in using surrogate markers in clinical trials. In *The Evaluation of Surrogate Endpoint*, pp. 13–51. Burzykowski, T., Molenberghs, G., and Buyse, M. (eds.). Springer, New York.

Chang, M. (2007). *Adaptive Design Theory and Implementation Using SAS and R.* Chapman & Hall/CRC, Taylor & Francis, Boca Raton, FL.

Chang, M. (2010). *Monte Carlo Simulations for the Pharmaceutical Industry.* Taylor & Francis/CRC, Boca, Raton. FL.

Chang, M. (2011). *Modern Issues and Methods in Biostatistics.* Springer, New York.

Chang, M. (2012). *Paradoxes in Scientific Inference.* CRC Press, Taylor & Francis Group, Boca Raton, FL.

Chow, S.C. and Chang, M. (2006). *Adaptive Design Methods in Clinical Trials.* Chapman & Hall/CRC, Boca Raton, FL.

Christakis, N. A. and Fowler, J. H. (2010), Social network sensors for early detection of contagious outbreaks, *PLoS ONE* 5(9): e12948.

Creed, B. (2009). *Darwin's Screens: Evolutionary Aesthetics, Time and Sexual Display in the Cinema.* Melbourne University Press, Carlton, Victoria, Australia.

Darwin, C. (1859). *On the Origin of Species by Means of Natural Selection, or the Preservation of Favoured Races in the Struggle for Life* (1st Ed.). London: John Murray.

Dennett, D. (1995). *Darwin's Dangerous Idea: Evolution and the Meanings of Life.* Touchstone, New York.

Dorigo, M. and Stützle, T. (2004). *Ant Colony Optimization.* MIT Press, Cambridge, MA.

Easley, D. and Kleinberg, J. (2010). *Networks, Crowds, and Markets—Reasoning about a Highly Connected World.* Cambridge University Press, Cambridge, NY.

Freedman, D.H. (November 2010) Lies, damned lies, and medical science. *The Atlantic.* www.theatlantic.com/magazine.archive/2010/01/lies-demand-lies-and-medicalscience/308269/

Freedman, D.H. (2010). *Wrong: Why Experts Keep Failing Us.* Little, Brown and Company, New York.

Erickson, G.J., Baldwin, J.W., and Varney, J.W. (1994). Application of the three-dimensional structures of protein target molecules in structure-based drug design. *Journal of Medicinal Chemistry* 37(8): 1035–54.

Frost, N. et al. (2010). Pluralism in qualitative research: The impact of different researchers and qualitative approaches on the analysis of qualitative data. *Qualitative Research.* 10(4): 01–20.

Gauch, H.G. (2003). *Scientific Method in Practice.* Cambridge University Press, Cambridge, UK.

Gintis, H. (2009). *Game Theory Evolving—A Problem-Centered Introduction to Modeling Strategy Interation* (2nd Ed.). Princeton University Press, Princeton, NJ.

Gintis, H. (2009). *The Bounds of Reason—Game Theory and the Unification of the Behavioral Sciences.* Princeton University Press, Princeton, NJ.

Gnedin, Sasha (2011). The Mondee Gills game. *The Mathematical Intelligencer.* http://www.springerlink.com/content/8402812734520774/fulltext.pdf

Goldhaber, A.S. and Nieto, M.M (2010). Photon and graviton mass limits. *Rev. Mod. Phys.* 82: 939–979.

Goodman, S. and Greenland, S. (2007a). *Assessing the Unreliability of the Medical Literature: A Response to "Why Most Published Research Findings Are False."* Johns Hopkins University, Department of Biostatistics, Baltimore, MD.

Goodman, S. and Greenland, S. (2007b). Why most published research findings are false: Problems in the analysis. *PLoS Medicine* 4(4): e168.

Gorroochurn, P. (2012). *Classic Problems of Probability.* John Wiley and Sons, New York.

Gubernator, K. and Böhm, H.-J., (Eds.). (1998). *Structure-Based Ligand Design (Methods and Principles in Medicinal Chemistry).* Weinheim: Wiley-VCH, New York.

Hadnagy, C. (2011). *Social Engineering.* Wiley Publishing Inc., Indianapolis, IN.

Hand, D.J. (2008). *Statistics: A Very Short Introduction.* Oxford University Press, Oxford, UK.

Havil, J. (2008). *Surprising Impossible: Solutions to Counterintuitive Conundrums.* Princeton University Press, Princeton, NJ.

Haw, M. (2005, January). Einstein's random walk. *Physics world,* pp. 19–22.

Herbranson, W.T., and Schroeder J. (2010). Are birds smarter than mathematicians? Pigeons (Columba livia) perform optimally on a version of the Monty Hall Dilemma. *J. Comp. Psychol.* 124(1):1–13.

Hilbe, J. M. (1977), *Fundamentals of Conceptual Analysis.* Kendall/Hunt Pub Co., Dubuque, IA.

Hill, A. B. (1965). The environment and disease: Association or causation? *Proceedings of the Royal Society of Medicine* 58: 295-300.

Holley, R. A.; Liggett, T. M. (1975). Ergodic theorems for weakly interacting infinite systems and the voter model. *The Annals of Probability* 3(4): 643–663.

Höltje, H-D., Sippl, W., Rognan, D., and Folkers, G. (2008). *Molecular Modeling: Basic Principles and Applications, 3rd Ed.* Wiley-VCH, New York.

Hotz, R.L. (2007, September). Most science studies appear to be tainted by sloppy analysis. *Science Journal.* online.wjs.com/news/articles/SB118972683557627104.

Howard, R.A (1960). *Dynamic Programming and Markov Processes.* The MIT. Press, Cambridge, MA.

Hume, D. (2011). *An Enquiry Concerning Human Understanding.* CreateSpace Independent Publishing Platform, Amazon.

Huxley, J. (2010). *Evolution: The Modern Synthesis.* The MIT Press, Cambridge, MA.

Huxley, T.H. (1863). *The Method of Scientific Investigation.*

Iba, H. Akiba, S., Higuchi, T., and Sato, T. (1992): *BUGS: A Bug-Based Search Strategy Using Genetic Algorithms.* PPSN, Brussels, Belgium, September 28–30.

Ioannidis, J.P.A. (2005a). Contradicted and Initially Stronger Effects in Highly Cited Clinical Research. *JAMA* 294(2): 218–228.

Ioannidis, J.P.A. (2005b). Why most published research findings are false. *PLoS Medicine* 2(8): e124.

Ioannidis, J.P.A. (2007). Why most published research findings are false: Author's reply to Goodman and Greenland. *PLoS Medicine* 4(6): e215.

Iverson, K.E. (2013). Notation as a Tool of Thought. http://www.jsoftware.com/papers/tot.htm, accessed on Nov. 2013.

Jackson, M.O. (2008). *Social and Economic Networks.* Princeton University Press, Princeton. NJ.

Jelle, J.G. and Aldo, S. (2011). Multiple testing for exploratory research. *Statistical Science* 26(4): 584–597.

Johnson, C.A. (2012). *The Information Diet: A Case for Conscious Consumption.* O'Reilly Media, Sebastopol, CA.

Kaptchuk T.J., Friedlander E., Kelley J.M., et al. (2010). Placebos without deception: A randomized controlled trial in irritable bowel syndrome. *PLoS ONE* 5(12): e15591.

Kennedy, J. and Eberhart, R. (1995). Particle swarm optimization, *Proceedings of the 1995 IEEE International Conference on Neural Networks*, pp. 1942–1948.

Kirk, R.E. (1995). *Experimental Design: Procedures for the Behavior Sciences* (3rd Ed.). Brooks/Cole Publishing Company, Pacific Grove, CA.

Kleinberg, J.M. (1998). Authoritative sources in a hyperlinked environment. Proc. 9th ACM-SIAM Symposium on Discrete Algorithms, San Francisco, CA, January 25–27.

Kuhn, T. (1977). *The Essential Tension.* University of Chicago Press, Chicago.

Levin, M. (1984). What Kind of Explanation is Truth? In Leplin, J. (Ed.). *Scientific Realism.* Berkeley: University of California Press, pp. 124–139.

Li, Y., Ng, K.C., Murray-Smith, D.J., et al. (1996). Genetic algorithm automated approach to design of sliding mode control systems. *Int J Control* 63(4): 721–739.

Lieberman, E., Hauert, C., and Nowak, M.A. (2005). Evolutionary dynamics on graphs. *Nature* 433(7023): 312–316

Luckhurst, R. (2005). *Late Victorian Gothic Tales*. Oxford University Press, Oxford, UK.

Materi, W. and Wishart, D.S. (2007). Computational systems biology in drug discovery and development: Methods and applications. *Drug Discov Today*. 12(7–8):295–303.

Mates, B. (1981). *Skeptical Essays*. University Chicago Press, Chicago, IL.

Matthews, M.R. (1994). *Science Teaching: The Role of History and Philosophy of Science*. Routledge, London.

Mayr, E. (1980). Some Thoughts on the History of the Evolutionary Synthesis. *In The Evolutionary Synthesis*, Mayr, E. and Provine, W.B. (Eds.). Harvard University Press. Cambridge, MA.

McCarthy, N. (2012). *Engineering: Beginner's Guides*. Oneworld Publication, Oxford, England.

McGrayne, S.B. (2012). *The Theory That Would Not Die: How Bayes' Rule Cracked the Enigma Code, Hunted Down Russian Submarines, and Emerged Triumphant from Two Centuries of Controversy*. Yale University Press, New Haven, CT.

Michail, K. (2009). Optimised Configuration of Sensing Elements for Control and Fault Tolerance Applied to an Electro-Magnetic Suspension, PhD Thesis, Loughborough University, UK

Miller, P. (2010). *The Smart Swarm*. The Penguim Group. New York.

Mitchell, M. (2009). *Complexity: A Guided Tour*. Oxford University Press, New York.

Moerman, D.E. and Jonas, W.B. (2002). Deconstructing the placebo effect and finding the meaning response. *Ann. Intern. Med.* 136(6): 471–476.

Moore, G. (2002). *Nietzsche, Biology and Metaphor*. Cambridge University Press, Cambridge, UK.

Moore, J.R. (2002). *History, Humanity and Evolution: Essays for John C. Greene*. Cambridge University Press, Cambridge, UK.

Motoda, H. and Ohara, K. (2009). Apriori, In Wu, X. and Kumar, V., (Eds.). *The Top Ten Algorithms in Data Mining*. Chapman & Hall/CRC, Boca Raton, FL, pp. 61–92.

Nash, J. (1951). Non-cooperative games. *Annals of Mathematics* 5:286–295.

Newton, Isaac (1687, 1713, 1726), *Philosophiae Naturalis Principia Mathematica*, University of California Press, ISBN 0-520-08817-4 , Third edition. P.794-796. From I. Bernard Cohen and Anne Whitman's 1999 translation.

Nowak, M.A. (2006). *Evolutionary Dynamics—Exploring the Equations of Life*. The Belknap Press of Harvard University Press, Cambridge, MA.

Nucci, Z.D. (2008). PhD thesis, Universitat Duisburg-Essen. Germany.

Parrondo, J. M. R. and Dinis, L. (2004). Brownian motion and gambling: From ratchets to paradoxical games. *Contemporary Physics* 45(2):147–157.

Pavlov, I.P. (1927/1960). *Conditional Reflexes*. Dover Publications, (New York the 1960 edition is not an unaltered republication of the 1927 translation by Oxford University Press http://psychclassics.yorku.ca/Pavlov/).

Pearson, Karl; Lee, A.; Bramley-Moore, L. (1899). Genetic (reproductive) selection: Inheritance of fertility in man. *Philosophical Translations of the Royal Statistical Society*, Ser. A 173: 534–539.

Petri, C.A. 1962. Kommunikation mit Automaten. English Translation, 1966: Communication with Automata. Technical Report RADC-TR-65-377, Rome Air Dev. Center, New York.

Pinski, G. and Narin, F. (1976). Citation influence for journal aggregates of scientific publications: Theory, with application to the literature of physics. *Information Processing & Management* 12(5): 297–312.

The *PLoS Medicine* Editors (2005), Minimizing mistakes and embracing uncertainty. *PLoS Med.* 2(8): e272.

Poli, R., Langdon, W.R., and McPhee, N.F. (2008). *A Field Guide to Genetic Programming*. Creative Commons Attribution. UK: England & Wales License.

Popper, K. (2002). The Logic of Scientific Discovery. Routledge Classics, New York.

Rescorla, R.A. (1967). Pavlovian conditioning and its proper control procedures. *Psychological Review* 74: 71–80

Rescorla, R.A. (1988) *American Psychologist*, 43: 151-160.

Rothman, K.J and Greenland, S (2005). Causation and causal inference in epidemiology. *Am. J. Publ. Health* 95:S144–150.

Russell, W.M.S. and Burch, R.L. (1959). *The Principles of Humane Experimental Technique*. Methuen, London.

Sakamoto, K. et al. (2007). A structural similarity evaluation by SimScore in a teratogenicity information sharing system. *J. Comput. Chem. Jpn.* 6(2): 117–122 (2007).

Salmon, W. C. (1967). *The Foundamental of Scientific Inference*. University of Pittsburgh Press, Pittsburgh PA.

Savage, L.J. (1954). *The foundations of statistics*. John Wiley & Sons, Inc., New York.

Schelling, T. (2006). *Micromotives and Macrobehavior*. W.W. Norton & Company, Inc. New York.

Schneider, G. and Baringhaus, K.H. (2008). *Molecular Design: Concepts and Applications*. Wiley-VCH Verlag, Frankfurt, Gemerny.

Schwartz, B. (2004). *The Paradox of Choice*. Harpercollins, New York.

Scott, J.G. and Berger, J.O. (2010). Bayes and empirical-Bayes multiplicity adjustment in the variable-selection problem. *Annuals of Statistics* 38: 2587–2619.

Selvin, S. (1975a). On the Monty Hall problem (letter to the editor). *American Statistician* 29(3): 134

Selvin, S. (1975b). A problem in probability (letter to the editor). *American Statistician* 29(1): 67

Shamoo, A. and Resnik, D. (2009). *Responsible Conduct of Research* (2nd Ed.). Oxford University Press, New York.

Shettleworth, Sara J. (2010). *Cognition, Evolution, and Behavior* (2nd Ed.). Oxford Univ. Press, UK.

Shubik, M. and Shapley, L. (1954). A method for evaluating the distribution of power in a committee system. *American Political Science Review* 48(3): 787–792.

Simpson, E.H. (1951). The interpretation of interaction in contingency tables. *Journal of the Royal Statistical Society*, Ser. B 13: 238–241.

Swift, A. (2006). *Political Philosophy, A Beginner's Guide for Students and Politicians*. (2nd Ed.). Polity Press, Cambridge, MA.

Székely, G.J. (1986). *Paradoxes in Probability Theory and Mathematical Statistics (Mathematics and Its Applications)*.

Szirtes, T. (2007). *Applied Dimensional Aanalysis and Modeling.* (2nd Ed.). Elsevier Inc., Burington, MA.

Tufte, E.R. (1983). *The Visual Display of Quantitative Information.* Graphics Press, Cheshire, CT.

van Batenburg, F.H., Gultyaev, A.P., and Pleij, C.W. (1995). An APL-programmed genetic algorithm for the prediction of RNA secondary structure. *Journal of Theoretical Biology* 174(3): 269–280.

Vidakovic, B. (2008). Bayesian Statistics Class handout. Brhttp://www2.isye.gatech.edu/~brani/isyebayes/handouts.html

Vincent, T.L and Brown, J.S. (2005). *Evolutionary Game Theory, Nature Selection, and Darwinian Dynamics.* Cambridge University Press. Cambridge, UK.

Wang, D. and Bakhai, A. (2006). Clinical Trials: A Practical Guide to Design, Analysis, and Reporting. London, UK.

Weisberg, H.J. (2012). *Bias and Causation—Models and Judgment for Valid Comparisons.* John Wiley and Sons, Hoboken, NJ.

West, G.B., Brown, J.H., and Enquist, B.J. (1999). The fourth dimension of life: Fractal geometry and allometric scaling of organisms. *Science* 284(5420): 1677–1679.

Willard, D.E. (2001), Self-verifying axiom systems, the incompleteness theorem and related reflection principles, *Journal of Symbolic Logic,* 66(2):536–596.

Willett, P. (1995). Genetic algorithms in molecular recognition and design. *Trends in Biotechnology* 13(12): 516–521.

Wigner, E.P. (1960). The unreasonable effectiveness of mathematics in the natural sciences. Richard Courant lecture in mathematical sciences delivered at New York University, May 11, 1959. *Communications on Pure and Applied Mathematics* 13: 1–14.

Wolpert, D. H., and Benford, G. (2010). What does Newcomb's paradox teach us? Cornell University. http://arxiv.org/abs/1003.1343

Wong, K.C., Peng, C., Wong, M.H., and Leung, K.S. (2011): Generalizing and learning protein-DNA binding sequence representations by an evolutionary algorithm. *Soft Computing* 15:1631–1642.

Wong, K., Leung, K., and Wong, M. (2010). Protein structure prediction on a lattice model via multimodal optimization techniques. *GECCO* 2010: 155-162.

Woodcock, J. (2005). FDA introductory comments: Clinical studies design and evaluation issues. *Clinical Trials* 2: 273-75

Wren, J.D. (2005) Truth, probability, and frameworks. *PLoS Med* 2(11): e361.

Yule, G.U. (1903). Notes on the theory of association of attributes in statistics. *Biometrika* 2(2): 121–134.

Zuckerman, E. W., and Jost, J.T. (2001). What makes you think you're so popular? Self-evaluation maintenance and the subjective side of the "friendship paradox." *Social Psychology Quarterly* 64(3): 207–223.

Index

T - #0373 - 101024 - C2 - 234/156/14 - PB - 9781482238099 - Gloss Lamination